IMPLEMENTATION OF HYPOPLASTICITY FOR FAST LAGRANGIAN SIMULATIONS

ADVANCES IN GEOTECHNICAL ENGINEERING AND
TUNNELLING

10

General editor:

D. Kolymbas

University of Innsbruck, Institute of Geotechnics and Tunnel Engineering

8. D. Kolymbas, ed. (2003), *Rational Tunnelling, Summerschool, Innsbruck 2003*, 428 pages, ISBN 3-8325-0350-1

9. D. Kolymbas, ed. (2004), *Fractals in Geotechnical Engineering, Exploratory Workshop, Innsbruck, 2003*, 174 pages, ISBN 3-8325-0583-0

10. P. Tanseng (2005), *Implementation of Hypoplasticity for Fast Lagrangian Simulations*, in print.

11. A. Laudahn (2005), *An Approach to 1g Modelling in Geotechnical Engineering with Soiltron*, in print.

Implementation of Hypoplasticity for Fast Lagrangian Simulations

Pornpot Tanseng
University of Innsbruck, Institute of Geotechnics and Tunnelling

E-mail: pornpot@ccs.sut.ac.th
Homepage: http://www.sut.ac.th/engineering/Civil/faculty/pornpot/

The first three volumes have been published by Balkema
and can be ordered from:

A.A. Balkema Publishers
P.O.Box 1675
NL-3000 BR Rotterdam
e-mail: orders@swets.nl
website: www.balkema.nl

Titelbild:
Chaloem Ratchamongkhon subway line (Bangkok)
The first subway project in Thailand

Bibliographic information published by Die Deutsche Bibliothek

Die Deutsche Bibliothek lists this publication in the Deutsche National-
bibliografie; detailed bibliographic data is available in the Internet at
http://dnb.ddb.de.

ISBN 3-8325-1073-7

ISSN 1566-6182

Logos Verlag Berlin
Comeniushof, Gubener Str. 47,
10243 Berlin
Tel.: +49 030 42 85 10 90
Fax: +49 030 42 85 10 92
INTERNET: http://www.logos-verlag.de

Abstract

The hypoplastic constitutive equations by Wu with and without structure tensor and by von Wolffersdorff were implemented in the finite element with explicit integration program FLAC. The influence of material nonlinearity, mass density and damping factor on numerical solutions with the explicit method were studied and reported. The implementation of hypoplasticity by Wu in FLAC was selected to simulate some laboratory tests i.e. biaxial test and simple shear test. The implementation was also used to simulate some typical geotechnical problems i.e. a spread footing on cohesionless soil, a trapdoor problem and an unlined circular tunnel. The results obtained with hypoplasticity agreed well with analytical solutions and experimental results. The hypoplastic constitutive equations and an elasto-plastic one with Mohr-Coulomb failure criterion were used to simulate normally consolidated Bangkok Clay. The hypoplasticity by Wu was selected for modelling of shield tunnelling in Bangkok Clay because the model gives good predictions of the undrained behavior. The earth pressure coefficient at rest K_0 used in the shield tunnelling simulations with FLAC is higher than the K_0 obtained from the Jaky's formula. The shield tunnelling was modelled by using actual dimensions from the construction records to generate the mesh. The surface settlements from the simulations are close to the observations. The empirical method by Peck underpredicts the surface settlements from observations.

Contents

Chapter 1

Introduction

Tunnel construction in soft soil deposits, especially soft clay, is a challenging work for geotechnical engineers. The tunnel route must often pass underneath urban areas where existing structures are located. In the past, engineers used empirical methods to estimate the influence of tunnelling to structures. The methods are solely based on the past experiences which may not be applicable if modern cross sections or new excavation procedures are introduced. Nowadays numerical methods are extensively used by engineers to estimate not only ground surface settlements, but also the stresses distribution in the soil.

One of the important keys in numerical simulation of tunnelling is the constitutive model. There are several models available for soft clay. Some of them are very simple to implement. Some of them are very complicate and require many parameters which are not easy to determine. This study uses a constitutive law which is simple in mathematical formulation and calibration to model normally consolidated clay. The constitutive law is implemented into a numerical program. The implemented code is verified and then it is used for simulations of shield tunnelling in normally consolidated clay.

1.1 Objective

General objectives of this study are:

- to implement hypoplastic constitutive equations into the numerical program FLAC and to verify the implemented code,

- to use hypoplastic constitutive equations to model normally consolidated (NC) Bangkok Clay,

- to develop a calibration method to obtain parameters for hypoplasticity from standard laboratory tests data of normally consolidated Bangkok Clay,

- to use the implemented constitutive equations for the simulation of shield tunnelling in Bangkok Clay.

1.2   Outline of the study

In this study, Chapter 2 describes the theory and background of the explicit finite difference program "FLAC". The program uses an element based algorithm; unknowns of the system are determined by exploiting the equation of motion. To use FLAC in solving quasi-static problems, the motion must be damped. The influences of damping, material nonlinearity and mass density are studied and documented. In chapter 3, some selected versions of hypoplasticity are implemented into FLAC. The implementation results are verified with "exact" solutions obtained using Euler forward integration. In Chapter 4, the implemented hypoplastic constitutive equation is used to simulate laboratory tests such as biaxial test and simple shear test. In addition, the implemented code is used to simulate some geotechnical problems, such as a spread footing, a trapdoor problem and a circular tunnel. An elasto-plastic Mohr-Coulomb model is also used to simulate these problems for comparison. Chapter 5 compares the common behavior of normally consolidated clay from experiments with the behavior predicted by hypoplasticity. After that, the hypoplastic constitutive equations are used to model normally consolidated Bangkok Clay. The calibration procedure for the model is described. In chapter 6, FLAC is used to simulate earth pressure balance (EPB) shield tunnelling in Bangkok subsoils. The implemented code of hypoplastic constitutive equation and the parameters calibrated from the experimental results for normally consolidated Bangkok clay are used as input for the simulations. The simulation results are then compared with monitored data collected during the construction. Finally, conclusions are drawn and recommendations for future works are included at the end of this study.

Chapter 2

Background of the explicit method

2.1 Introduction

This chapter describes the background of the explicit dynamic method used to solve the set of differential equations with given initial and boundary conditions. In this study, the program FLAC (Fast Lagrangian Analysis of Continua) version 4.0 (17) is used for numerical simulations. In the explicit finite difference method, every derivative in the set of governing equations is converted into an algebraic expression written in terms of the undefined field variables (e.g. stress or displacement) at discrete points in space. FLAC uses an explicit time integration method to solve the algebraic equations generated at each step. Even if FLAC is used to find a static solution, the dynamic equations of motion are used to ensure that the numerical scheme is stable even if the situation being modelled is unstable. With nonlinear materials, there is always the possibility of physical instability – e.g., the sudden collapse of a pillar, which can be modelled with FLAC efficiently. Figure 2.1 shows the calculation procedure used in FLAC. First, the equations of motion are used to calculate the new velocities and displacements from stresses, accelerations and forces. Next, the new velocities are used to calculate strain increments, and a constitutive equation is invoked to derive new stresses from the strain increments. The main idea is that

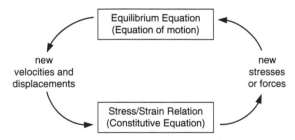

Figure 2.1: Basic explicit calculation cycle in $FLAC$ (17)

the calculated "wave speed" always keeps ahead of the actual wave speed, so that the equations always operate on known values which are constant for the duration of each calculation step. The advantage of this procedure is that no iteration process is required in computing stresses from strain increments in an element, even when the

3

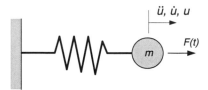

Figure 2.2: Schematic representation of mass-spring system (17)

constitutive law is highly nonlinear. On the contrary, in an implicit method (which is commonly used in finite element programs), several cycles of iteration are required before equilibrium and compatibility are obtained.

As FLAC uses the time integration method to obtain field variables, there is no need to formulate a global stiffness matrix. Therefore, updating of all coordinates at each time increment is quite simple. The incremental displacements are added to the current coordinates; therefore, the grid moves and deforms with the material it represents. This formulation is termed a "Lagrangian", in contrast to "Eulerian" formulation, in which the material moves and deforms relative to a stationary grid. A small-strain framework is used in the constitutive formulation at each step; however, over many steps, it is equivalent to a large-strain framework. The mathematical formulation of the explicit method is described in detail in the following sections.

2.2 Continuum field equations

FLAC invokes the equations of motion, constitutive relations, initial and boundary conditions to solve a boundary value problem. This section reviews the basic governing differential equations for a deformable body.

2.2.1 The equation of motion

A mass fixed with a spring is the simplest system used in describing the motion of a system (Fig. 2.2). The system consists of a mass m connected to a spring fixed at one end; the force $F(t)$ is applied on the mass and causes motion which can be described in terms of acceleration $\ddot{u}$, velocity $\dot{u}$ and displacement u. The governing equation is

$$m\ddot{u} = \sum F \qquad (2.1)$$

If several forces act on the mass, the equilibrium is reached when the acceleration becomes zero (i.e. the summation of all forces acting on the mass is zero, $\sum F = 0$). This law of motion is also used in FLAC to solve static problems. In continuum mechanics, Equation 2.1 is written in generalized form as

$$\rho \ddot{u}_i = \frac{\partial \sigma_{ij}}{\partial x_j} + \rho g_i \qquad (2.2)$$

where ρ is mass density, $\ddot{u}_i$ is the nodal acceleration vector, x_i is the coordinate vector, g_i is the gravitational acceleration vector, and σ_{ij} is the stress tensor.

2.2.2 General constitutive relation

The relationship of stress and strain of a deformable body is represented by using set of equations known as constitutive law. The strain rate is obtained from the velocity gradient as follows:

$$\dot{e}_{ij} = \frac{1}{2} \left[\frac{\partial \dot{u}_i}{\partial x_j} + \frac{\partial \dot{u}_j}{\partial x_i} \right] \qquad (2.3)$$

where $\dot{e}_{ij}$ is the strain-rate vector; $\dot{u}_i$ is the velocity. The stress rate can be calculated from the strain-rate by applying a constitutive equation which is of the functional form

$$\dot{\sigma}_{ij} := M(\sigma_{ij}, \dot{e}_{ij}, \kappa) \qquad (2.4)$$

where κ is a history parameter which may or may not be presented depending on the particular law. Generally, non-linear constitutive laws are written in incremental or rate form because the relationship between stress and strain is not unique. The stresses at a current time increment are calculated by adding the current stress increments to the previously calculated stresses. Implementation of constitutive equations into FLAC is described in detail in Chapter 3. Furthermore, in large-strain mode, FLAC adjusts the stress tensor due to the finite rotation during a time increment by

$$\sigma_{ij} := \sigma_{ij} + (\omega_{ik}\sigma_{kj} - \sigma_{ik}\omega_{kj})\Delta t \qquad (2.5)$$

where

$$\omega_{ij} = \frac{1}{2} \left\{ \frac{\partial \dot{u}_i}{\partial x_j} - \frac{\partial \dot{u}_j}{\partial x_i} \right\} \qquad (2.6)$$

Equation 2.5 is used before the constitutive equation (Eq.2.4) is applied. In small-strain mode, stress adjustments due to finite strain components are not made.

2.3   Discrete numerical formulations

The continuum field equations are converted into the finite difference form which can
be integrated explicitly through time. The state at the end of the increment is based
solely on the displacements, velocities and accelerations at the beginning of the increment. To acquire accurate results, the time increments must be adequately small
so that the accelerations are approximately constant during an increment. Because
the time increment must be adequately small, simulations typically require many
thousands of increments. This section describes the formulation of the element, the
finite difference equations and the damping used to obtain quasi-static solutions.

2.3.1   Discretization and finite element formulation

In FLAC2D, a continuum problem is discretized into a finite element mesh which
is composed of quadrilateral zones. Generally, each zone is internally subdivided
into two overlaid sets of constant-strain triangular elements (elements a, b, c and d)
as shown in Figure 2.3. Each triangular sub-element has four stress components

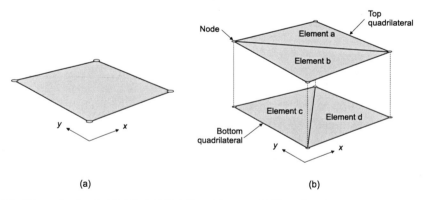

Figure 2.3: Discretization in FLAC; (a) FLAC's grid zone; (b) FLAC's grid zone subdivided into two
sets of constant-strain triangular elements

($\sigma_{xx}, \sigma_{yy}, \sigma_{zz}$, and σ_{xy}); therefore, 16 stress components have to be stored for each
quadrilateral zone. The use of triangular elements eliminates the problem of hour-
glass deformation which may occur when constant-strain quadrilateral elements are
used. Using plane-strain or axisymmetric analyses often overpredicts collapse loads
because a kinematic restraint in the out-of-plane direction is introduced. To over-
come this problem, the isotropic stress and strain components are taken to be constant over the whole quadrilateral zone, while the deviatoric components are treated
separately for each triangular sub-element (17).

2.3.2 Finite difference equations

For a constant-strain triangular element, a finite difference can be written as

$$\left\langle \frac{\partial f}{\partial x_i} \right\rangle = \frac{1}{A} \sum_{s=1}^{3} \langle f \rangle n_i \Delta s \tag{2.7}$$

where Δs is the length of a side of the triangle; $\langle f \rangle$ is the average value of f (scalar, vector or tensor) over the side of the element; A is the area of the element; x_i is the position vector; and n_i is the unit outward normal vector of the surface s.

By using Equation 2.7, strain rates can be written in terms of nodal velocities for

$$\frac{\partial \dot{u}_i}{\partial x_j} \cong \frac{1}{2A} \sum_{s=1}^{3} \left(\dot{u}_i^{(a)} + \dot{u}_i^{(b)} \right) n_j \Delta s \tag{2.8}$$

$$\dot{e}_{ij} = \frac{1}{2} \left[\frac{\partial \dot{u}_i}{\partial x_j} + \frac{\partial \dot{u}_j}{\partial x_i} \right] \tag{2.9}$$

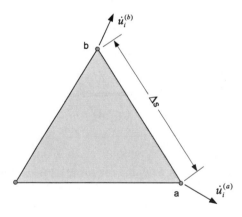

Figure 2.4: Typical triangular element with velocity vectors (17)

2.3.3 Nodal forces

The net nodal force vector $\sum F_i$ is the sum of the nodal forces from the element stresses, nodal forces due to gravity, and nodal forces from boundary stresses. The net nodal force is used to obtain the acceleration from Newton's second law of motion.

After the stresses have been calculated from the strain rates by applying a constitutive equation, the nodal forces corresponding to these stresses are determined. As

FLAC uses a constant-strain element for the element discretisation, the stresses in
each element are also constant. The tractions on each side of the triangular elements
correspond to the stresses in the element. The traction along the side of the triangle
must be converted into equivalent nodal forces ($F^{(1)}$ and $F^{(2)}$), as illustrated in Figure 2.5. It can be seen that each node receives two forces contributed from the two

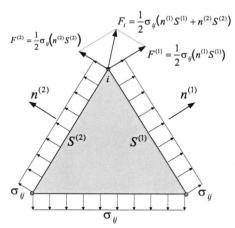

Figure 2.5: Determination of nodal force vector

adjoining sides. Hence,

$$F_i = \frac{1}{2}\sigma_{ij}\left(n^{(1)}S^{(1)} + n^{(2)}S^{(2)}\right) \tag{2.10}$$

where $n^{(1)}$ and $n^{(2)}$ are the unit normal to the surfaces and $S^{(1)}$ and $S^{(2)}$ are the
length of the boundary segments. As each FLAC's grid zone consists of two sets of
two triangular elements (Fig. 2.3), the nodal forces from triangular elements which
share the same node are accumulated. The accumulated nodal forces are then averaged to obtain a nodal force for a quadrilateral's node.

The body forces F_i^g play an important role in most geotechnical problems. The
forces must be applied at the node of the element; therefore, the calculation of
equivalent nodal forces is required. The equivalent nodal force from gravity can
be calculated from

$$F_i^g = g_i m_g \tag{2.11}$$

where m_g is the lumped gravitational mass at the node; for a triangular element,
which shares the same node, contributes one-third of its mass to the node.

The stress boundary conditions have to be converted into equivalent nodal forces. In
FLAC, at a stress boundary, equivalent nodal forces are calculated from

$$F_i^b = \sigma_{ij}^{(b)} n_j \Delta s \tag{2.12}$$

where n_i is the unit outward normal vector of the boundary segment, and Δs is the length of the boundary segment on which the boundary stress is applied for plane strain problems.

2.3.4 Time integration for the equations of motion

From the equations of motion (Eq. 2.2), if the body is in equilibrium ($\sum F_i = 0$) or in steady-state flow (i.e. plastic flow), the acceleration of the node becomes zero. If the net force on the node is not equal to zero, then the node is accelerated; the induced acceleration can be calculated as follows:

$$\ddot{u}_i|_{(t)} = \frac{\sum F_i|_{(t)}}{m} \tag{2.13}$$

The accelerations are integrated through time using the central difference rule to obtain velocity increments. After that, the calculated increments are added to the middle of the previous time increment to determine the velocities at the next half time increment as follows:

$$\dot{u}_i|_{(t+\Delta t/2)} = \dot{u}_i|_{(t-\Delta t/2)} + \ddot{u}_i|_{(t)}\Delta t \tag{2.14}$$

The displacements at the end of the increment are obtained by integrating the velocities through time and adding the results to the displacements at the beginning of the increment:

$$u_i|_{(t+\Delta t)} = u_i|_{(t)} + \dot{u}_i|_{(t+\Delta t/2)}\Delta t \tag{2.15}$$

It should be noted that the acceleration and the velocity are assumed as constant during the time integration cycle; therefore, the time increment must be adequately small to obtain accurate results.

In large-strain calculation (Lagrangian analysis), the coordinates of grid points are updated by adding incremental displacements to the grid points at the beginning of the increment:

$$x_i|_{(t+\Delta t)} = x_i|_{(t)} + u_i|_{(t+\Delta t)} \tag{2.16}$$

2.4 Time incrementation and stability

In the explicit method, the time integration is done through an increment of time, Δt. The incremental quantities obtained from the integration are added to the quantities at the beginning of the increment. The time increment must be sufficiently small so that the state quantities (i.e. stresses, velocites, and displacements) can be advanced and still remain an accurate representation of the problem. The largest time increment which gives a stable calculation can be called "critical time increment" or "stability limit". If the time increment is larger than the critical one, numerical instability may occur and may lead to an unbounded solution. Generally, the exact critical time increment is not easy to determine; therefore, an estimation of the critical time increment is used. In addition, some factors of safety must be applied to the estimated critical time increment to prevent numerical instability. The critical time increment for an elastic continuum, discretized into elements of size Δx, is

$$\Delta t = \frac{\Delta x}{c_p} \tag{2.17}$$

where c_p is the p-wave speed of the material:

$$c_p = \sqrt{\frac{D}{\rho}}. \tag{2.18}$$

D is the confined modulus which can be written in terms of the bulk modulus K, and the shear modulus G, of the material as follows:

$$D = K + 4G/3 \tag{2.19}$$

For using the explicit method for solving a static problem, a fictitious nodal mass is used as relaxation factor which is adjusted for optimum speed of convergence. In FLAC, the time increment is set to 1.0 and the nodal masses are adjusted to obtain the critical time increment. Generally, a safety factor of 0.5 is applied to the critical time increment to ensure numerical stability (Itasca (17)). For a triangular element of area A, which has the minimum propagation distance of $A/\Delta x_{\text{max}}$, the nodal mass of each grid point can be calculated as follows:

$$m_n = \sum \frac{D(\Delta x_{\text{max}})^2}{6A} \tag{2.20}$$

It should be noted that the grid point mass is taken as one-third of the element mass. Practically, it is advantageous to keep the element size as large as possible, because the critical time step is approximately proportional to the minimum propagation distance. However, for accurate results, a fine mesh is often required. The best way is to have a mesh that is as uniform as possible, since the critical time increment is based on the smallest element dimension in the model.

The effect of structural elements and interface elements is incorporated by adding the equivalent masses to the summation of Equation 2.20, assuming $\Delta t = 1.0$. Each structural element connected to a grid node contributes an extra mass to the summation as follows:

$$m_{\text{struct}} = 4k, \tag{2.21}$$

where k is the diagonal term corresponding to the structural node. A safety factor of 4.0 is applied to Equation 2.21 because a system of connected elements may have a faster oscillation than a single element.

2.5 Mechanical damping

FLAC uses the explicit finite difference scheme (as discussed in section 2.3) to solve the equations of motion. If static solutions are required, the motion must be damped. Generally, damping is added to a model to limit numerical oscillations or to express physical damping. Cundall (4) proposed a form of damping where the damping force on a node is proportional to the magnitude of the out-of-balance force. The sign of the damping force is always opposite to the out-of-balance force to ensure that vibration modes are damped and energy is always dissipated,

$$F_d \propto |F| \mathrm{sgn}(\dot{F}), \qquad (2.22)$$

where F is the nodal out-of-balance force. If the system is in a steady-state condition the damping force vanishes. Equation 2.22 can be written in terms of velocity at the node, the out-of-balance force, and a damping constant as follows:

$$(F_i)_\mathrm{d} = \alpha \left| \sum F_i^{(t)} \right| \mathrm{sgn}\left(\dot{u}_i^{(t-\Delta t/2)} \right) \qquad (2.23)$$

where α is the damping constant, which is dimensionless and does not depend on properties or boundary conditions (default value is 0.8 in FLAC). Equation 2.23 is incorporated into equation 2.14:

$$\dot{u}_i|_{(t+\Delta t/2)} = \dot{u}_i|_{(t-\Delta t/2)} + \left\{ \sum F_i^{(t)} - (F_d)_i \right\} \frac{\Delta t}{m_n} \qquad (2.24)$$

where m_n is a fictitious nodal mass which is derived from Equation 2.20. The influence of damping is illustrated by an example in Chapter 2.6.

2.6 Simulation of a one-dimensional quasi-static problem

The explicit method may be illustrated with a simple example, a 1D rod. The geometry of the rod is shown in Figure2.6. In this example, the influence of material nonlinearity, mass density and damping factor on the numerical solutions is demonstrated.

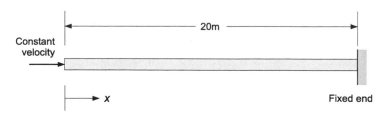

Figure 2.6: Geometry and boundary conditions of a rod for an unconfined compression test

2.6.1 Finite element discretization

The rod is discretized into 20 constant-strain elements as shown in Figure 2.7. The rod is fixed at the right end by setting the nodal velocity to zero; a constant velocity is applied at the left end. The velocities and displacements are calculated at nodes while stresses are calculated at the elements. Three constitutive models are used in the simulations: linear elasticity, hyperbolic elasticity and hypoplasticity. The incremental forms of the one-dimensional equations for these models are as follows:
Linear elastic constitutive equation:

$$\Delta\sigma = E \cdot \Delta\varepsilon \tag{2.25}$$

Hyperbolic elastic constitutive equation:

$$\Delta\sigma = E \cdot \Delta\varepsilon \cdot \frac{(\sigma_{\max} - \sigma)^2}{\sigma_{\max}^2} \tag{2.26}$$

Hypoplastic constitutive equation (54):

$$\Delta\sigma = E \cdot \Delta\varepsilon - E \cdot \frac{\sigma}{\sigma_{\max}}|\Delta\varepsilon| \tag{2.27}$$

where E is the Young's modulus and $\sigma_{\max}$ is the stress at limit state. The parameters of the constitutive models are shown in Table 2.1

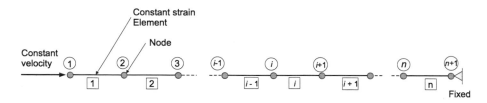

Figure 2.7: The rod is discretized into $n = 20$ elements

Constitutive model	Parameters
Linear elasticity	$E = 400$ N/m^2
	$\rho = 100$ kg/m^3
Hyperbolic elasticity	$E = 400$ N/m^2
	$\sigma_{\max} = 2$ N/m^2
	$\rho = 100$ kg/m^3
Hypoplasticity	$E = 400$ N/m^2
	$\sigma_{\max} = 2$ N/m^2
	$\rho = 100$ kg/m^3

Table 2.1: Parameters for the models used in the unconfined compression-decompresion test simulation

2.6.2 The explicit equations for one-dimensional problems

The equation of motion in a three-dimensional form with body forces (Eq. 2.2) can be re-written in a one-dimensional form as follows

$$\frac{\partial \sigma}{\partial x} = \rho \frac{\partial^2 u}{\partial t^2} + \rho g_x \tag{2.28}$$

Equation 2.28 can be rearranged to obtain the acceleration as follows

$$\ddot{u} = \frac{1}{\rho} \cdot \frac{\partial \sigma}{\partial x} + g_x \tag{2.29}$$

If the quasi-static condition is required by using the equation of motion (Eq. 2.29), damping must be introduced to the formulations. Accelerations at nodal points are used to obtain the magnitude of damping. In addition, the damping acceleration (or damping force) is proportional to the acceleration, and the sign of the damping is always opposite to the current velocity. The damped acceleration for a one-dimensional problem as proposed in FLAC (17) is

$$\ddot{u}_{\text{damped}} = \ddot{u}_x^i - \alpha |\ddot{u}_x^i| \text{sgn}(\dot{u}_x^i) \tag{2.30}$$

where $\ddot{u}_x^i$ and $\dot{u}_x^i$ are the acceleration and velocity at the node i, respectively. From equation 2.30, it can be seen that the acceleration at the node is always reduced by the amount of $\alpha |\ddot{u}_x^i|$. The system is in equilibrium if the acceleration becomes zero. Numerically, absolute zero acceleration can not be achieved; therefore, a limit value must be defined.

The accelerations at the nodal points are used to obtain the velocities by time integration

$$\dot{u}_x^i(t + \frac{\Delta t}{2}) = \dot{u}_x^i(t - \frac{\Delta t}{2}) + \ddot{u}_{\text{damped}} \Delta t \tag{2.31}$$

The displacements at the nodal points are calculated from the velocities

$$u_{(t+\Delta t)}^i = u_{(t)}^i + \dot{u}_{(t+\frac{\Delta t}{2})}^i \cdot \Delta t \tag{2.32}$$

Figure 2.8 shows a flowchart for the numerical procedure of the explicit method used to obtain a quasi-static solution. This method has been implemented in a MATLAB program. The implemented source code is given in Appendix A.

2.6.3   Influence of material nonlinearity

The material model supplies a stable time increment through its effect on wave speed. Wave speed is always constant in a linear material; therefore, the stable time increment depends then solely on the smallest dimension of the elements in the system. In a nonlinear material, for instance hyperbolic elasticity or hypoplasticity, the wave speed changes as the stiffness of the material changes. The effect of material nonlinearity on the solution was studied by implementing three constitutive equations, namely elasticity, hyperbolic elasticity, and hypoplasticity in the EFDM. The implemented code is used to simulate an unconfined compression test of a rod as described in section 2.6.1. The rod was loaded until the axial strain reaches 4%. During loading, node number 10 and element number 10 are used to observe the acceleration and stress-strain response, respectively. The results are compared with the "exact" solution obtained by using the Euler forward integration.

Figure 2.9, 2.10, and 2.11 show the simulation results obtained with the elastic model, the hyperbolic elastic model and the hypoplastic model, respectively. From the figures, the stress-strain curves of the three models are similar to the exact results obtained by Euler forward integration method. In case of the linear elastic model, the accelerations at the observed point show oscillation around zero (Fig.2.9b); the maximum magnitude of the acceleration is 8×10^{-3} m/s^2. For the hyperbolic elastic model (Fig.2.10b), the accelerations at the observed point shows a peak of 3.19×10^{-4} m/s^2; after that, the nodal accelerations converge to zero without any oscillation. In case of the hypoplastic model (Fig. 2.11b), the acceleration at the observing point shows a peak of 5.16×10^{-4} m/s^2, then small oscillations appear before the acceleration converges to zero. The simulation results show that the material model does not have any influence on the results obtained by using the explicit finite difference method. The stress-strain curves of the linear elastic material, the nonlinear elastic material and the hypoplastic material show no difference from the "exact" solutions. The solutions are quasi-static, as the accelerations of the node converge to zero, i.e. $\sum F = 0$. For linear elasticity, the oscillation at the observed point appears because the stiffness of the material is always constant. The steady state solution is obtained because a damping (Eq.2.30) is added to the system; moreover, the inertia of the system impedes the accelerations. In contrast, the nodal accelerations of the hyperbolic elasticity show no oscillations, as the stiffness of the model decreases with proceeding deformation; thus, it retards the accelerations of the system. For hypoplasticity, the system shows slightly underdamped behavior. However, this small oscillation does not have any effect on the result.

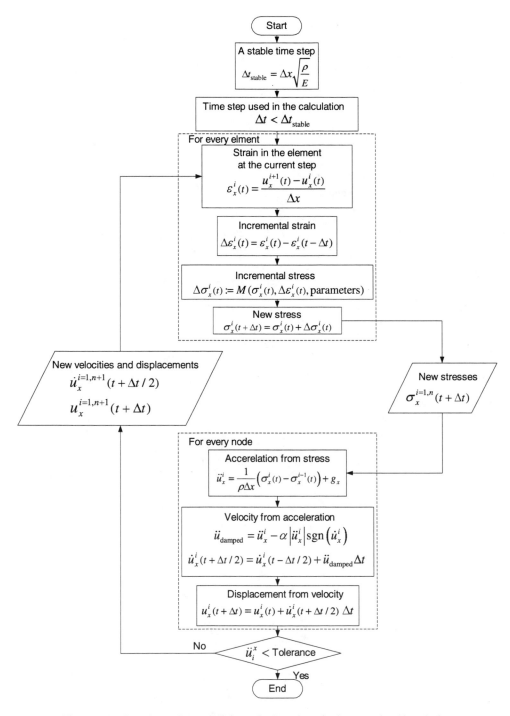

Figure 2.8: Flowchart of the explicit method used to obtain a quasi-static solution

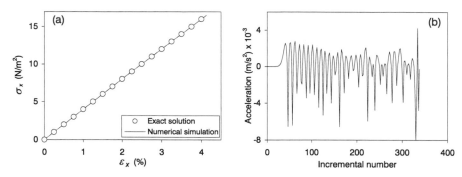

Figure 2.9: Simulation results obtained with the linear elastic model: (a) Stress-strain curves at element no. 10; (b) Acceleration at node number 10

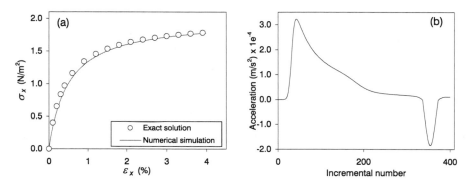

Figure 2.10: Simulation results obtained by using the hyperbolic elastic model: (a) Stress-strain curves at element no. 10; (b) Acceleration at node number 10

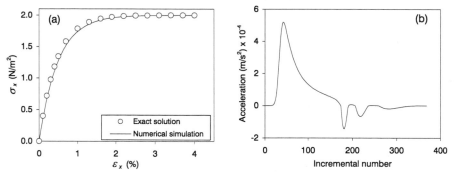

Figure 2.11: Simulation results obtained by using the hypoplastic model: (a) Stress-strain curves at element no. 10; (b) Acceleration at node number 10

In conclusion, the material nonlinearity has influence on the accelerations of the system but does not have any influence on the simulation results in terms of stress-strain curves. Material with nonlinearity provides damping to the system, so the accelerations converge to zero with small oscillations. In contrast, linear elastic materials provide no damping to the system; therefore, a fictitious (artificial) damping must be included in the formulations to accelerate the convergence. In addition, the nodal mass also affects the numerical solution. The influence of nodal mass (mass density) is described in section 2.6.4.

2.6.4 Influence of mass density

As shown in Figure 2.8, before the time integration loop begins, the stable time increment must be determined. Furthermore, the time increment Δt used in the time integration must be adequately smaller than the stable time increment which is calculated by

$$(\Delta t)_{\text{stable}} = \frac{\Delta x}{v_c} \tag{2.33}$$

where Δx is the element length. The speed of the sound in the material is calculated from

$$v_c = \sqrt{\frac{E}{\rho}} \tag{2.34}$$

where E is Young's modulus of the material, and ρ is its mass density. The critical time increment depends, thus, on the stiffness and the density of the material. For example, if two materials have similar density, the material with high stiffness also has higher wave speed than the material with low stiffness. The stable time increment depends on the wave speed of the material: the lower the wave speed, the higher the stable time increment.

To demonstrate the influence of nodal masses on the numerical performance, the simulations of the unconfined compression test (as described in section 2.6.1) were performed by using the explicit finite difference method implemented into a MAT-LAB program. The rod was modelled by using hypoplasticity and the simulations were performed by using three values of mass density: 100 kg/m^3, 500 kg/m^3 and 1000 kg/m^3. The rod is loaded in compression until the axial strain reaches 4%. A node at the middle length of the rod is used to observe the accelerations during loading. An element at the middle length of the rod is also used to observe the stress-strain response due to loading. The results of the simulations are then compared to the exact solution which was obtained with the Euler forward integration method.

The obtained stress-strain curves of every test reach a maximum stress, σ_{max}, at 2 N/m^2 and the shape of the curves are similar to the exact solution (Fig. 2.12). Figure 2.13 shows the plots of accelerations at the observed points versus incremental number for three simulations with different mass densities. The acceleration reaches peaks of $5.16 \times 10^{-4} \text{ m/s}^2$, $1.05 \times 10^{-4} \text{ m/s}^2$, and $0.5 \times 10^{-4} \text{ m/s}^2$ for the mass densities of 100 kg/m^3, 500 kg/m^3, and 1000 kg/m^3, respectively. After the peak acceleration, small oscillations appear before the acceleration converges to zero.

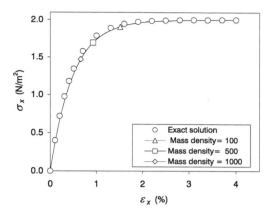

Figure 2.12: Influence of nodal mass on the numerical solutions: hypoplastic constitutive model; damping factor = 0.8

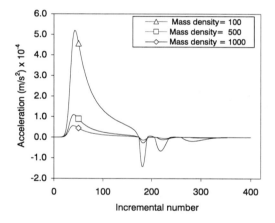

Figure 2.13: Influence of nodal mass on the numerical solutions: hypoplastic constitutive model; damping factor = 0.8

Figure 2.14 shows the peak accelerations at the observed node versus the mass densities. The plot shows that the peak accelerations reduce nonlinearly as the mass densities increase; the peak accelerations are almost constant if the mass densities are greater than 2000 kg/m^3. It can be seen that the magnitudes of accelerations and oscillations at the node can be reduced by increasing the nodal mass. Moreover, increasing the nodal mass does not affect the numerical solution. Generally, the nodal mass used for obtaining quasi-static problem can be a fictitious mass, which is in contrast to full dynamic problems where the real nodal masses must be used (Itasca (17)). In addition, increasing of the mass density also increases the critical time increment; as a result, the computation time is reduced.

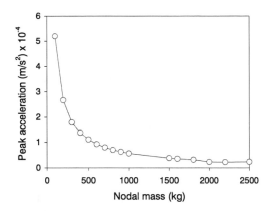

Figure 2.14: Relationship between nodal mass and peak acceleration (out-of-balance force) : hypoplastic constitutive model; damping factor = 0.8

2.6.5 Influence of damping factor

To obtain quasi-static solutions from the explicit integration method, an appropriate value of the damping factor must be applied to Equation 2.30 for 1D problems or Equation 2.23 for 3D problems. A series of unconfined compression tests with different damping factors (0, 0.4, and 0.8) was performed to investigate the influence of the damping factor on the numerical solution. The boundary conditions of the test are described in section 2.6.1. The explicit finite difference method implemented in a MATLAB program (as shown in Appendix A) was used for these simulations.

A hypoplastic constitutive model was used to model the rod with the parameters given in Table 2.1. The rod was compressed until the axial strain 4%. A nodal point at middle of the rod was used as an observation point for the accelerations. The stress-strain responses were also observed at an element at the middle of the rod. Then the results were compared to the "exact" solution obtained by using the Euler forward integration method.

Figure 2.15(a), 2.16(a) and 2.17(a) show the plots of the deviatoric stresses versus strains for damping factors of 0.0, 0.5, and 0.8 respectively. Figure2.15(b), 2.16(b), and 2.17(b) show the plots of the accelerations at the observing point versus incremental number.

The results show that if no damping is applied to the equation of motion, oscillations appear and the simulation results deviate from the "exact" solution. In addition, the accelerations at the observed point severely oscillate before converge to zero. As the accelerations converge to zero, the deviatoric stresses also converge to the "exact" solution. In case of the damping factor of 0.5, a similar behavior as in the case of no damping is observed. For the damping factor of 0.8, the stress-stress curve from the simulation is similar to the "exact" solution. Moreover, the oscillations of the observed node reduce significantly. Therefore, the damping factor used for hypoplasticity the damping factor should not be zero. In the following chapters the

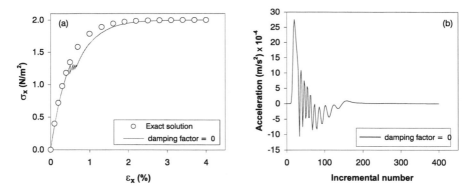

Figure 2.15: Damping factor, $\alpha = 0.0$; (a) Stress-strain; (b) Acceleration versus increment number

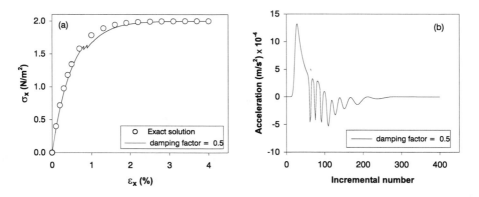

Figure 2.16: Damping factor, $\alpha = 0.5$; (a) Stress-strain; (b) Acceleration versus increment number

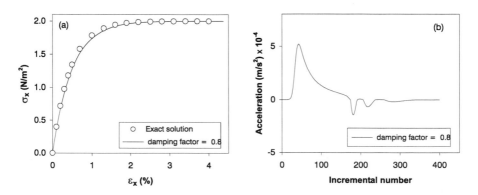

Figure 2.17: Damping factor, $\alpha = 0.8$; (a) Stress-strain; (b) Acceleration versus increment number

damping factor of 0.8 is used for all calculations.

It can be seen from the example that the explicit method is quite simple. The quasi-static results obtained from the explicit method are close to the "exact" solutions. The implementation of a highly non-linear, for examplehyperbolic elastic model or hypoplastic model, is a straight-forward process. The simulations with an elastic model or a plastic model require similar computational effort. Although iterations are not required in the method like the finite element method, the time increment must be smaller than the critical time increment, which means that large number of steps must be used to reach equilibrium. The purpose of the damping factor in the explicit method is to make the accelerations become zero ($\partial \sigma / \partial x = 0$) unlike the full dynamic analyzes in which a physically correct damping factor should be applied.

In summary, if an appropriate mass density is used and a suitable damping factor is applied, correct quasi-static solutions can be obtained. Fictitious mass can be added to the nodes to speed up the calculations. Although the added mass is unrealistic, it does not have any effect on the numerical solution. Moreover, a suitable damping factor can be used to speed up the convergence.

Chapter 3

Implementation of hypoplasticity in FLAC

This chapter describes a numerical implementation of hypoplastic constitutive equations. In FLAC, a constitutive equation is programmed in C++ and then compiled as DLL (Dynamic Link Library) file. The pre-release Version 4.1 of FLAC requires a DLL file, which "must be" compiled by Microsoft Visual C++ (17). The source codes mentioned in this research were compiled with Microsoft Visual C++, Version 6.0. The main task of the implemented DLL file is to calculate and return a new set of stress components from the given strain increments and the old set of stress components. Moreover, the model also provides other information, such as names of the variables. After the constitutive equation has been implemented, the one-zone quadrilateral grid composed of 4 triangular sub-elements as mentioned in section 2.3 is used for verifying the implementation results. Some simple strain paths are given to the grid and responses of the model are observed. Furthermore, the results obtained from the one-zone grid are compared with the known results from the forward integration method.

The constitutive models can also be implemented in FLAC by using a programming language embedded within FLAC, called "FISH". The code is implemented in a separate file and contains special statements and references to special variables that correspond to local entities within a single zone. A user-defined constitutive model implemented in FISH runs directly without any compilation. However, the execution time of a model implemented in FISH is typically longer than the model which is implemented and compiled by Visual C++.

The implemented codes are also used to simulate a spread footing problem, the trapdoor problem and a circular tunnel. After that, the results are compared with a Mohr-Coulomb model which is built-in in FLAC.

3.1 Framework of hypoplasticity

The framework of hypoplasticity is described by Kolymbas (19) and Wu et al. (53). A hypoplastic constitutive equation is defined by assuming a tensorial function **H**, such that

$$\overset{\circ}{\sigma} = \mathbf{H}(\sigma, \dot{\varepsilon}) \tag{3.1}$$

where $\dot{\varepsilon}$ is the stretching (strain rate) and $\overset{\circ}{\sigma}$ is the co-rotational (JAUMANN) stress rate, which is defined as follows

$$\overset{\circ}{\sigma} = \dot{\sigma} + \sigma\dot{\omega} - \dot{\omega}\sigma \tag{3.2}$$

where $\dot{\sigma}$ is the time derivative of the CAUCHY stress σ, and $\dot{\omega}$ is the rotation rate (spin tensor) which is the antimetric part of the velocity gradient grad $\mathbf{v}$, i.e. (grad $\mathbf{v} -$ (grad $\mathbf{v})^{\mathrm{T}})/2$. Another assumption is that the function $\mathbf{H}$ in Equation 3.1 is continuously differentiable for all $\dot{\varepsilon}$ except at $\dot{\varepsilon} = 0$.

To formulate more concrete constitutive equations from the general formulation, some restrictions are imposed on the function $\mathbf{H}$

- The function $\mathbf{H}$ must be positively homogeneous of the first order in $\dot{\varepsilon}$ to describe rate independent behavior

$$\mathbf{H}(\sigma, \lambda\dot{\varepsilon}) = \lambda\mathbf{H}(\sigma, \dot{\varepsilon}), \tag{3.3}$$

  where λ is a positive scalar factor.

- The function $\mathbf{H}$ should be homogeneous in σ to incorporate the behavior of pressure sensitivity

$$\mathbf{H}(\lambda\sigma, \dot{\varepsilon}) = \lambda^n\mathbf{H}(\sigma, \dot{\varepsilon}), \tag{3.4}$$

  where λ is a scalar factor and n is the order of homogeneity in stress. Equation 3.4 implies that the tangential stiffness is proportional to the n-th power of the stress level $(\mathrm{tr}\sigma)^n$. In some hypoplastic version, the order of homogeneity assumed to be unity; therefore, the experimental results which are conducted under different stress level can be normalized by $(\mathrm{tr}\sigma)$.

- The function must fulfil the condition of objectivity

$$\mathbf{H}(\mathbf{Q}\sigma\mathbf{Q}^{\mathrm{T}}, \mathbf{Q}\dot{\varepsilon}\mathbf{Q}^{\mathrm{T}}) = \mathbf{Q}\mathbf{H}(\sigma, \dot{\varepsilon})\mathbf{Q}^{\mathrm{T}} \tag{3.5}$$

  where $\mathbf{Q}$ is an orthogonal tensor. The representation theorem for a tensorial function of two symmetric tensors σ and ε can be written as follows

$$\begin{aligned}
\overset{\circ}{\sigma} = {}& \phi_0\mathbf{1} + \phi_1\sigma + \phi_2\dot{\varepsilon} + \phi_3\sigma^2 + \phi_4\dot{\varepsilon}^2 + \phi_5(\sigma\dot{\varepsilon} + \dot{\varepsilon}\sigma) \\
& + \phi_6(\sigma^2\dot{\varepsilon} + \dot{\varepsilon}\sigma^2) + \phi_7(\sigma\dot{\varepsilon}^2 + \dot{\varepsilon}^2\sigma) + \phi_8(\sigma^2\dot{\varepsilon}^2 + \dot{\varepsilon}^2\sigma^2)
\end{aligned} \tag{3.6}$$

where $\mathbf{1}$ is the unit tensor. The coefficients $\phi_i (i = 0 \dots 8)$ are scalar functions of invariants and joint invariants of σ and $\dot{\varepsilon}$:

$$\begin{aligned}
\phi_i = {}& \phi_i(\mathrm{tr}\sigma, \mathrm{tr}\sigma^2, \mathrm{tr}\sigma^3, \mathrm{tr}\dot{\varepsilon}, \mathrm{tr}\dot{\varepsilon}^2\mathrm{tr}\dot{\varepsilon}^3, \\
& \mathrm{tr}(\sigma\dot{\varepsilon}), \mathrm{tr}(\sigma^2\dot{\varepsilon}), \mathrm{tr}(\sigma\dot{\varepsilon}^2), \mathrm{tr}(\sigma^2\dot{\varepsilon}^2))
\end{aligned} \tag{3.7}$$

The constitutive equation can be decomposed into two parts representing the reversible and irreversible behavior of the material (54)

$$\overset{\circ}{\boldsymbol{\sigma}} = \mathbf{L}(\boldsymbol{\sigma}, \dot{\boldsymbol{\varepsilon}}) + \mathbf{N}(\boldsymbol{\sigma}, \dot{\boldsymbol{\varepsilon}}) \tag{3.8}$$

where $\mathbf{L}$ is linear in $\dot{\varepsilon}$ and $\mathbf{N}$ is nonlinear in $\dot{\varepsilon}$. Furthermore, a generalized hypoplastic equation assumes the form (54)

$$\overset{\circ}{\boldsymbol{\sigma}} = \mathbf{L}(\boldsymbol{\sigma}, \dot{\boldsymbol{\varepsilon}}) + \mathbf{N}(\boldsymbol{\sigma}) \|\dot{\boldsymbol{\varepsilon}}\| \tag{3.9}$$

where $\|\dot{\boldsymbol{\varepsilon}}\| = \sqrt{\mathrm{tr}(\dot{\varepsilon}^2)}$ is the Euclidean norm. It can be seen that the relationship of stress rates and strain rates can be described by using Equation 3.9 alone without any predefined yield surface and plastic potential. Moreover, the decomposition in elastic and plastic parts is not used in developing the constitutive equation. Compared to the plasticity theory, hypoplastic constitutive models are incrementally nonlinear; moreover, Equation 3.9 can be used for loading and unloading.

3.2 Implementation of a 1D hypoplastic constitutive equation

The one-dimensional hypoplastic constitutive equation is the simplest form used to illustrate the implementation of the equation in FLAC. The uniaxial relationship between stress increment and strain increment by Wu et al. (54) (Eq. 2.27) is used. The equation is implemented in FISH as shown in Appendix B. The code is verified by using it to simulate an 1D compression test. The problem is discretized to a one-zone mesh as shown in Figure 3.1.

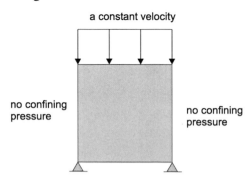

Figure 3.1: Element for triaxial element test

The zero initial axial stress is applied to the mesh and no lateral confining pressure is applied to the model. The compression is simulated by applying a constant velocity of -1×10^{-4} m/step at the top boundary of the mesh. The axial compressive strain is applied up to 4% strain, then the axial compressive strain is reduced until the axial stress becomes zero. Figure 3.2 shows the plots of the stress-strain curve from the simulation and the exact solution obtained from the Euler forward integration method with sufficiently small time increment.

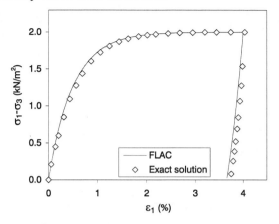

Figure 3.2: Numerical simulation of unconfined compression test with one-dimensional hypoplasticity: $E = 400$ kN/m^2, $\sigma_{max} = 2$ kN/m^2, velocity $= 1 \times 10^{-4}$m/step

The result obtained from FLAC with the implemented hypoplasticity is similar to the exact solution; therefore, 1D hypoplasticity can be successfully implemented in FLAC.

3.3 Implementation of hypoplasticity by Wu

Previous section describes the implementation of 1D hypoplasticity in FLAC. In this section the implementation of hypoplasticity by Wu et al. (53) is described.

$$\overset{\circ}{\sigma} = C_1 \text{tr}(\sigma)\dot{\varepsilon} + C_2 \frac{\text{tr}(\sigma\dot{\varepsilon})}{\text{tr}\sigma}\sigma + C_3 \frac{\sqrt{\text{tr}(\dot{\varepsilon}^2)}}{\text{tr}\sigma}\sigma^2 + C_4 \frac{\sqrt{\text{tr}(\dot{\varepsilon}^2)}}{\text{tr}\sigma}\sigma^{*2} \qquad (3.10)$$

The equation consists of four tensorial terms; each term is multiplied with a material parameter (C_1, C_2, C_3 and C_4). The deviatoric stress σ^* is defined as

$$\sigma^* = \sigma - \frac{1}{3}(\text{tr}\sigma)\mathbf{1} \qquad (3.11)$$

To implement the hypoplastic constitutive equation in FLAC, the equation must be in incremental form. Therefore, the time increment Δt is multiplied with 3.10, which can be re-written in the incremental form as

$$\Delta\sigma = C_1(\text{tr}\sigma)\Delta\varepsilon + C_2 \frac{\text{tr}(\sigma\Delta\varepsilon)}{\text{tr}\sigma}\sigma + C_3 \frac{\sqrt{\text{tr}(\Delta\varepsilon^2)}}{\text{tr}\sigma}\sigma^2 + C_4 \frac{\sqrt{\text{tr}(\Delta\varepsilon^2)}}{\text{tr}\sigma}\sigma^{*2} \quad (3.12)$$

FLAC uses different coordinate notations for model implementation as shown in Figure 3.3. The basic formulation for FLAC is for a two-dimensional plane-strain

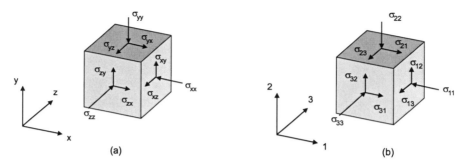

Figure 3.3: Coordinates and stresses notations: (a) For FLAC modelling; (b) For constitutive model implementation

model. For the plane-strain condition, the z-dimension of the problem is assumed to be very large, so the displacements in the x-y plane are independent of the z coordinate. Therefore, FLAC considers only the stress increments $\Delta\sigma_{xx}, \Delta\sigma_{yy}, \Delta\sigma_{zz}$ and $\Delta\sigma_{xy}$ in plane-strain analysis. The incremental expressions of hypoplastic constitutive equation which are implemented in FLAC are

$$\Delta\sigma_{11} = C_1(\mathrm{tr}\boldsymbol{\sigma})\Delta\varepsilon_{11} + C_2\frac{\mathrm{tr}(\boldsymbol{\sigma}\Delta\boldsymbol{\varepsilon})}{\mathrm{tr}\boldsymbol{\sigma}}\sigma_{11} + C_3\frac{\sqrt{(\Delta\varepsilon)^2}}{\mathrm{tr}\boldsymbol{\sigma}}\sigma_{11}^2 + C_4\frac{\sqrt{(\Delta\varepsilon)^2}}{\mathrm{tr}\boldsymbol{\sigma}}\sigma_{11}^{*\,2}$$

$$\Delta\sigma_{22} = C_1(\mathrm{tr}\boldsymbol{\sigma})\Delta\varepsilon_{22} + C_2\frac{\mathrm{tr}(\boldsymbol{\sigma}\Delta\boldsymbol{\varepsilon})}{\mathrm{tr}\boldsymbol{\sigma}}\sigma_{22} + C_3\frac{\sqrt{(\Delta\varepsilon)^2}}{\mathrm{tr}\boldsymbol{\sigma}}\sigma_{22}^2 + C_4\frac{\sqrt{(\Delta\varepsilon)^2}}{\mathrm{tr}\boldsymbol{\sigma}}\sigma_{22}^{*\,2}$$

$$(3.13)$$

$$\Delta\sigma_{33} = C_1(\mathrm{tr}\boldsymbol{\sigma})\Delta\varepsilon_{33} + C_2\frac{\mathrm{tr}(\boldsymbol{\sigma}\Delta\boldsymbol{\varepsilon})}{\mathrm{tr}\boldsymbol{\sigma}}\sigma_{33} + C_3\frac{\sqrt{(\Delta\varepsilon)^2}}{\mathrm{tr}\boldsymbol{\sigma}}\sigma_{33}^2 + C_4\frac{\sqrt{(\Delta\varepsilon)^2}}{\mathrm{tr}\boldsymbol{\sigma}}\sigma_{33}^{*\,2}$$

$$\Delta\sigma_{12} = C_1(\mathrm{tr}\boldsymbol{\sigma})\Delta\varepsilon_{12} + C_2\frac{\mathrm{tr}(\boldsymbol{\sigma}\Delta\boldsymbol{\varepsilon})}{\mathrm{tr}\boldsymbol{\sigma}}\sigma_{12} + C_3\frac{\sqrt{(\Delta\varepsilon)^2}}{\mathrm{tr}\boldsymbol{\sigma}}\sigma_{12}^2 + C_4\frac{\sqrt{(\Delta\varepsilon)^2}}{\mathrm{tr}\boldsymbol{\sigma}}\sigma_{12}^{*\,2}$$

Equations 3.13 are implemented in Visual C++ and compiled as DLL file. The implemented code is shown in Appendix B.

After the hypoplastic constitutive equation version of Wu was implemented, the code was verified by simulating a triaxial test with a one-zone mesh. The geometry for the triaxial test simulation is shown in Figure 3.4.

Axisymmetric formulation was used in the simulation. Horizontal fixities were applied at the symmetry axis. A constant confining pressure of 100 kN/m² was applied at the lateral boundary. Material parameters for hypoplastic constitutive model were taken from Kolymbas (19) as shown in Table 3.1. Loading on the sample was simulated by applying a constant velocity at the top of the mesh. The influence of velocity

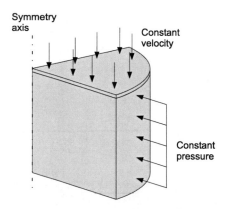

Figure 3.4: Geometry of a triaxial test simulation

on the numerical solution was also investigated. The magnitude of the applied velocities was varied for each test as shown in Table3.2. A forward integration method with adequately small time step was used to obtain an exact solution for comparison.

C_1	C_2	C_3	C_4
-106.5	-801.5	-797.1	1077.7

Table 3.1: Parameters for hypoplastic model (19)

Test no.	Simulation method	Applied velocity
1	FLAC	1×10^{-4} m/step
2	FLAC	2×10^{-4} m/step
3	FLAC	4×10^{-4} m/step
4	FLAC	5×10^{-4} m/step
5	Forward integration	1×10^{-6} m/m/step

Table 3.2: Applied velocities used in the simulations

Figure 3.5 shows the stress-strain curves obtained by using methods and applied velocities specified in Table 3.2. At the applied velocity of 1×10^{-4} m/step, the solution obtained with FLAC is similar to the solution obtained from forward integration method. The numerical results obtained from FLAC diverge from the exact solution if the magnitude of the applied velocity is larger than 1×10^{-4}. For the applied velocities of 4×10^{-4} m/step and 5×10^{-4} m/step, the deviatoric stresses significantly differ from the exact solution.

Additionally, the peak deviator stresses obtained from the simulations are plotted versus the applied velocity, as shown in Figure 3.6. The figure shows that, if the applied velocities are smaller than 2×10^{-4} m/step, the peak deviator stresses obtained from FLAC are equal to the exact peak deviator stresses. In contrast, if the

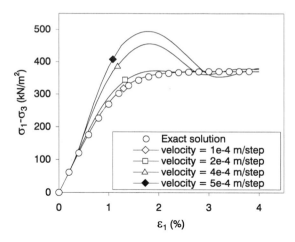

Figure 3.5: Influence of velocities on numerical solution

applied velocities are larger than 2×10^{-4} m/step, the peak deviator stresses immensely deviate from the exact peak deviator stress. Therefore, the applied velocity of 2×10^{-4} m/step is the critical applied velocity for a one-zone mesh simulation of a triaxial test. The applied velocity must be smaller than this value. However, the calculation time and the accuracy of the results must also be concerned in the applied velocity selections. Generally, velocity 10 times smaller than the critical velocity is sufficient for common problems.

An applied velocity of 1×10^{-6} m/step, which is 10 times smaller than the critical

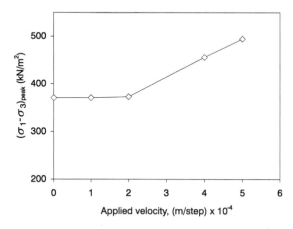

Figure 3.6: Plots of the peak deviator stresses versus the applied velocities

velocity, is used to simulate a triaxial test with loading and unloading. The results of the simulation are compared to the exact solution obtained from a forward integration method with an adequately small incremental step. Figure 3.7 (a) shows the plot of deviatoric stress versus axial strain and Figure 3.7 (b) shows the plot of volumetric

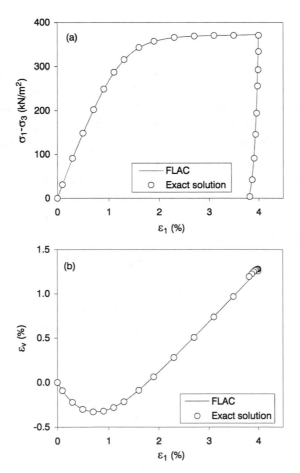

Figure 3.7: Numerical simulation results of a triaxial test; hypoplastic constitutive equation version Wu; (a) deviatoric stress versus axial strain; (b) volumetric strain versus axial strain

strain versus axial strain. From the plots, it can be seen that the simulation results obtained with FLAC are quite good compared to the exact solutions.

3.4 Implementation of hypoplasticity with a structure tensor

The hypoplastic constitutive equation was originally developed for non-cohesive soils. However, non-cohesive soils and cohesive soils have some common behaviors; therefore, the idea of extending the hypoplastic constitutive equation to cohesionless soils arose. This is achieved with the incorporation of a suitable structure tensor, which has a unit of stress, into the hypoplastic constitutive equation. After the structure tensor is introduced into the equation of Wu, the equation can be used to describe the cohesive soil with the same set of parameters for several stress states. The hypoplastic constitutive equation version Wu (Eq.3.10) is used for developing the extended version. A tensorial parameter $\mathbf{S}$ is introduced to Equation 3.10 to describe soil history and anisotropy. $\mathbf{S}$ expresses the internal attraction between the grains and depends on the loading history. The generalized form of the incremental structure stresses is as follows:

$$\Delta \mathbf{S} = \mathbf{g}(\sigma, \mathbf{S}, \Delta \varepsilon) \tag{3.14}$$

where $\mathbf{S}$ is the structure stress. Because $\mathbf{S}$ is a fictitious stress, it does not affect the equilibrium condition; therefore, it is also called 'internal stress' or 'back stress'. The evolution equation for the structure stress was proposed as follows:

$$\Delta \mathbf{S} = \mathrm{tr}\mathbf{D}\left(\mu \mathbf{1} + \lambda \frac{\sigma^*}{\mathrm{tr}\sigma}\right) \tag{3.15}$$

where μ and λ are material parameters. For isotropic compression the second term vanishes. However, for deviatoric stress paths the second term is activated automatically. $\mathbf{S}$ is incorporated into Equation 3.10 in the following way:

$$\Delta \sigma = C_1 \mathrm{tr}(\sigma + \mathbf{S})\Delta \varepsilon + C_2 \frac{\mathrm{tr}((\sigma + \mathbf{S})\Delta \varepsilon)}{\mathrm{tr}(\sigma + \mathbf{S})}(\sigma + \mathbf{S}) +$$
$$+ C_3 \frac{\sqrt{\mathrm{tr}\Delta \varepsilon^2}}{\mathrm{tr}(\sigma + \mathbf{S})}(\sigma - \mathbf{S})^2 + C_4 \frac{\sqrt{\mathrm{tr}\Delta \varepsilon^2}}{\mathrm{tr}(\sigma + \mathbf{S})}(\sigma^* - \mathbf{S}^*)^2 \tag{3.16}$$

where $\mathbf{S}^*$ is defined as

$$\mathbf{S}^* = \mathbf{S} - \frac{1}{3}(\mathrm{tr}\mathbf{S})\mathbf{1} \tag{3.17}$$

Equation 3.16 is implemented into FLAC by using a similar procedure as for the basic equation. For a plane deformation problem, only four stress components

($\sigma_{11}, \sigma_{22}, \sigma_{33}$ and σ_{12}) are required for the implementation. It must be noted that averaging of the structure stress components must be done as a quadrilateral zone of FLAC consists of four sub-elements. The stresses are stored for the whole quadrilateral zone but the user-written constitutive equation is called for each triangular sub-zone (2 or 4 elements). Therefore, the structure stresses must be accumulated and averaged to obtain a representative structural stress for a quadrilateral zone. The implemented codes in Visual C++ are shown in Appendix B.

After the hypoplastic constitutive equation with a structure tensor is implemented and compiled, a simulation of a triaxial test is done for verification of the code. A one-zone quadrilateral mesh is used to model the test; the test is assumed to be axisymmetric as described in section 3.3. The parameters for hypoplasticity with a structure tensor are shown in Table 3.3. The sample is assumed to be isotropically consolidated with 204.75 kN/m^2 with initial isotropic structure stress Q of 8.04 kN/m^2. The sample is loaded in compression until the axial strain in the sample reaches 30%. Then the sample is unloaded until the axial stress becomes zero. The results from the numerical simulation are compared with the exact results obtained from Euler forward integration with an adequately small time increment. Figure 3.8

C_1	C_2	C_3	C_4	$Q(\,$kN/m$^2)$	μ	λ
-8.45	-112.80	-105.19	94.39	8.04	100	20

Table 3.3: Parameters for hypoplasticity with structure tensor

(a) shows the plots of deviator stress versus axial strain and Figure 3.8 (b) shows the plots of volumetric strain versus axial strain of simulations from FLAC compared to the exact solutions. The simulated deviatoric stress curve is slightly lower than the "exact" solution (about 1.9%) and the simulated volumetric stress is slightly higher than the exact solution (at maximum of 6%). The deviation of the simulation results are due to the averaging scheme in FLAC.

3.5 Implementation of hypoplasticity of von Wolffersdorff

The hypoplastic constitutive model of von Wolffersdorff (51) was designed to model cohesionless soil at any densities with only eight parameters. The equation uses the current stress and the void ratio as state variables. The general incremental form of the equation is:

$$\Delta\sigma = \mathbf{F}(\sigma, e, \Delta\varepsilon) \tag{3.18}$$

where $\Delta\sigma$ is the stress response for a given strain increment $\Delta\varepsilon$ at one material point; σ is the actual stress, and e is the void ratio. The incremental void ratio Δe

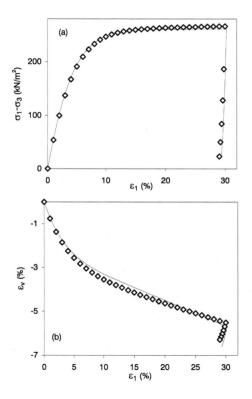

Figure 3.8: Numerical simulation results of a triaxial test; hypoplastic constitutive equation version von Wu with structure tensor; confining pressure = 204.75 kN/m^2; (a) deviatoric stress versus axial strain; (b) volumetric strain versus axial strain

relates to the volumetric strain increment $\mathrm{tr}(\Delta\varepsilon)$, i.e.:

$$\Delta e = (1+e)\mathrm{tr}(\Delta\varepsilon) \tag{3.19}$$

This constitutive law takes the influence of the pressure level and the density on the behavior of the soils into consideration. Figure 3.9 shows the plots of possible range of void ratios versus pressure, where e_i is the upper limit, e_d is the lower limit, and e_c is the critical state void ratio.

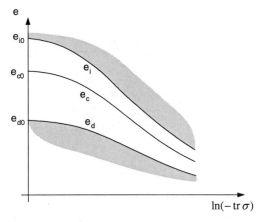

Figure 3.9: The range of possible void ratio as a function of stress; e_i is the upper limit void ratio, e_d is lower limit void ratio, and e_c is the critical void ratio (32)

It can be seen from the curves that, e_i, e_d, and e_c decrease as the mean stress increases; the law of compression is

$$\frac{e_i}{e_{i0}} = \frac{e_c}{e_{c0}} = \frac{e_d}{e_{d0}} = \exp\left[-\left(-\frac{\mathrm{tr}\sigma}{h_s}\right)^n\right] \tag{3.20}$$

This hypoplastic constitutive equation has been adapted to the yield function of Matsuoka/Nakai. The hypoplastic constitutive equation of von Wolffersdorff is given by the tensorial equation

$$\Delta\sigma = f_b f_e \frac{1}{\mathrm{tr}(\hat{\sigma}^2)}\left\{F^2\Delta\varepsilon + a^2\mathrm{tr}(\hat{\sigma}\Delta\varepsilon)\hat{\sigma} + f_d aF(\hat{\sigma}+\hat{\sigma}^*)\sqrt{\mathrm{tr}(\Delta\varepsilon^2)}\right\} \tag{3.21}$$

with:

$$a = \frac{\sqrt{3}(3-\sin\varphi_c)}{2\sqrt{2}\sin\varphi_c} \tag{3.22}$$

F is the stress function according to Matsuoka/Nakai

$$F = \sqrt{\frac{1}{8}\tan^2\psi + \frac{2-\tan^2\psi}{2+\sqrt{2}\tan\psi\cos 3\vartheta}} - \frac{1}{2\sqrt{2}}\tan\psi \tag{3.23}$$

where

$$\tan \psi = \sqrt{3\text{tr}(\hat{\boldsymbol{\sigma}}^2)} \tag{3.24}$$

and

$$\cos 3\vartheta = -\sqrt{6} \frac{\text{tr}(\hat{\boldsymbol{\sigma}}^* \cdot \hat{\boldsymbol{\sigma}}^* \cdot \hat{\boldsymbol{\sigma}}^*)}{[\text{tr}(\hat{\boldsymbol{\sigma}}^* \cdot \hat{\boldsymbol{\sigma}}^*)]^{3/2}} \tag{3.25}$$

The pyknotropy functions are

$$f_e = \left(\frac{e_c}{e}\right)^{\beta} \tag{3.26}$$

and

$$f_d = \left(\frac{e - e_d}{e_c - e_d}\right)^{\alpha} \tag{3.27}$$

And the barotropy function is

$$f_b = \tag{3.28}$$

$$\left(\frac{h_s}{n}\right)\left(\frac{1+e_i}{e_i}\right)\left(\frac{e_{i0}}{e_{c0}}\right)^{\beta}\left(-\frac{\text{tr}\boldsymbol{\sigma}}{h_s}\right)^{1-n}\left[3 + a^2 - \sqrt{3}a\left(\frac{e_{i0}-e_{d0}}{e_{c0}-e_{d0}}\right)^{\alpha}\right]^{-1}$$

where

$$\hat{\boldsymbol{\sigma}} = \frac{\boldsymbol{\sigma}}{\text{tr}\boldsymbol{\sigma}} \tag{3.29}$$

$$\hat{\boldsymbol{\sigma}}^* = \hat{\boldsymbol{\sigma}} - \frac{1}{3}\mathbf{1} \tag{3.30}$$

and void ratio increment is calculated as follows:

$$\Delta e = (1 + e)(\Delta\varepsilon_{11} + \Delta\varepsilon_{22} + \Delta\varepsilon_{33}) \tag{3.31}$$

Four stresses components are required in the implementation in two-dimension version of FLAC i.e.

$$\Delta\sigma_{11} = f_b f_e \frac{1}{\text{tr}(\hat{\boldsymbol{\sigma}}^2)}\left\{F^2\Delta\varepsilon_{11} + a^2\text{tr}(\hat{\boldsymbol{\sigma}}\Delta\boldsymbol{\varepsilon})\hat{\sigma}_{11} + f_d aF(\hat{\sigma}_{11} + \hat{\sigma}_{11}^*)\sqrt{\text{tr}(\Delta\boldsymbol{\varepsilon}^2)}\right\}$$

$$\Delta\sigma_{22} = f_b f_e \frac{1}{\text{tr}(\hat{\boldsymbol{\sigma}}^2)}\left\{F^2\Delta\varepsilon_{22} + a^2\text{tr}(\hat{\boldsymbol{\sigma}}\Delta\boldsymbol{\varepsilon})\hat{\sigma}_{22} + f_d aF(\hat{\sigma}_{22} + \hat{\sigma}_{22}^*)\sqrt{\text{tr}(\Delta\boldsymbol{\varepsilon}^2)}\right\}$$

$$\tag{3.32}$$

$$\Delta\sigma_{33} = f_b f_e \frac{1}{\text{tr}(\hat{\boldsymbol{\sigma}}^2)}\left\{F^2\Delta\varepsilon_{33} + a^2\text{tr}(\hat{\boldsymbol{\sigma}}\Delta\boldsymbol{\varepsilon})\hat{\sigma}_{33} + f_d aF(\hat{\sigma}_{33} + \hat{\sigma}_{33}^*)\sqrt{\text{tr}(\Delta\boldsymbol{\varepsilon}^2)}\right\}$$

$$\Delta\sigma_{12} = f_b f_e \frac{1}{\text{tr}(\hat{\boldsymbol{\sigma}}^2)}\left\{F^2\Delta\varepsilon_{12} + a^2\text{tr}(\hat{\boldsymbol{\sigma}}\Delta\boldsymbol{\varepsilon})\hat{\sigma}_{12} + f_d aF(\hat{\sigma}_{12} + \hat{\sigma}_{12}^*)\sqrt{\text{tr}(\Delta\boldsymbol{\varepsilon}^2)}\right\}$$

A set of equations (Eq. 3.32) is implemented into Visual C++ and compiled as DLL file. The implemented code is shown in Appendix B.

The implemented DLL file is used to simulate a triaxial compression test with a one-zone mesh. Here we consider axisymmetric problems. The parameters for the model are the granular hardness h_s, the critical friction angle φ_c, the void ratios e_{c0}, e_{d0} and e_{i0}, and the exponents n and β; their values are shown in Table 3.4. The sample is assumed to be consolidated isotropically at confining pressure of 100 kN/m^2 with an initial void ratio of 0.7. The sample is loaded in compression by applying a constant velocity of 1×10^{-6} m/step at the top of the model. Loading is performed until the axial strain in the sample reaches 16%; after that, the sample is unloaded until the deviatoric stress becomes zero. The numerical simulation results are compared with the exact solutions which obtained by using Euler's forward integration method with sufficient small time increments. Figure 3.10 (a) shows the plots of deviator

h_s (N/m^2)	n	φ_c	e_{c0}	e_{d0}	e_{i0}	α	β
1x10^9	0.29	32	0.91	0.61	1.09	0.19	2.00

Table 3.4: Parameters for hypoplastic model version von Wolffersdorff

stress versus axial strain for FLAC simulation and the "exact" solution. Figure 3.10 (b) shows the corresponding plots of volumetric strain versus axial strain. It can be concluded from the curves that the results from FLAC are very close to the "exact" solutions.

In summary, the implementing process of the hypoplastic constitutive equation in FLAC is straightforward. After the constitutive equation is implemented, it should be verified thoroughly on a one-zone mesh with simple boundary conditions such as triaxial test. The strain path is applied to the model, and the stress and volumetric strain responses are obtained. After that, the results are compared with the known solution; for example, the results which are obtained from Euler's forward method with adequately small time increments. The simulation with a one-zone mesh is an effective way to verify the implementation results. The applications of the implemented model in geotechnical simulations are presented in Chapter 4.

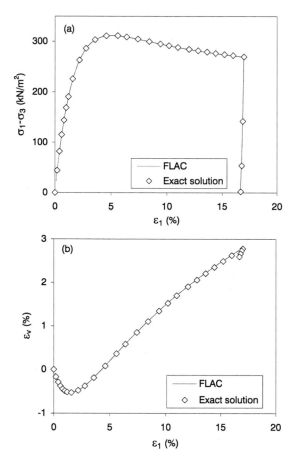

Figure 3.10: Numerical simulation results of a triaxial test; hypoplastic constitutive equation version von Wolffersdorff; (a) deviatoric stress versus axial strain; (b) volumetric strain versus axial strain

Chapter 4

Further numerical simulation examples

Once the hypoplastic constitutive equations have been implemented in FLAC and the results of the implementation have been verified, one of the implemented constitutive equations is selected to simulate some laboratory experiments and geotechnical problems. In this study, the hypoplastic constitutive equation by Wu is selected for the simulations. The reason for choosing this constitutive equation is discussed in Chapter 5.

4.1 Simulation of a biaxial test

4.1.1 Simulation procedure

The simulation of a biaxial test was performed by (16; 44; 6) with FEM. In this study, the biaxial test is simulated by FLAC with the same geometry as used by Hügel (16) but with different soil models. Figure 4.1 shows the geometry and the boundary conditions of the test. A soil sample of 40 mm width and 140 mm height is confined laterally by cell pressure of 200 kN/m^2. An initial stress of 200 kN/m^2 is assumed in the soil sample. Constant velocities of 2×10^{-7} m/step are applied at the top and bottom of the mesh. Non-homogeneity of the sample is simulated by introducing a low-strength zone of 20 mm $\times$ 20 mm in the sample as shown in the dashed area. The soil is discretized into small grid zones (as described in 2.3). Three meshes are used to observe the influence of mesh density as shown in Figure 4.2. The soil is modelled by hypoplasticity by Wu and Mohr-Coulomb (M-C) model. The parameters for hypoplasticity are taken from Kolymbas (19). The friction angle calculated from these parameters is 40.5°. For the low-strength zone, the parameters are calibrated by assuming the friction angle $\varphi = 37°$. The parameters of M-C model are calibrated from the simulation results of triaxial test with hypoplasticity. The parameters for both models are summarized in Table 4.1 and 4.2.

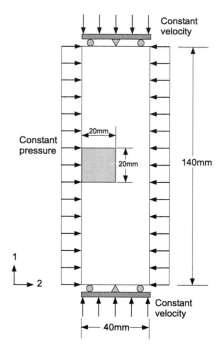

Figure 4.1: Geometry and boundary conditions for a biaxial test. The imperfection zone is hatched (16).

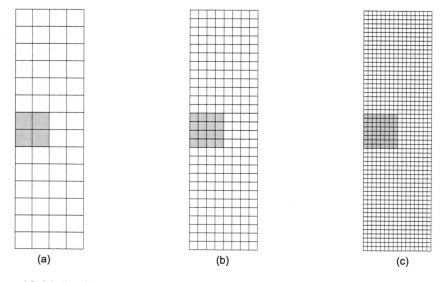

Figure 4.2: Meshes for biaxial test simulation; (a) Coarse mesh; (b) Medium mesh; (c) Fine mesh.

Parameter	Normal zone	Low-strength zone
C_1	-106.5	-106.5
C_2	-801.5	-886.9
C_3	-797.1	-882.2
C_4	1077.7	1446.1

Table 4.1: Hypoplastic parameters for the biaxial test. Parameters for normal zone is taken from Kolymbas (19); parameters for low strength zone are obtained by assuming $\varphi = 37°$

Parameter	Normal zone	Low-strength zone
φ	$40.5°$	$37.0°$
ψ	$13°$	$13°$
E_i	32 MN/m^2	32 MN/m^2
ν	0.3	0.3

Table 4.2: M-C parameters for the biaxial test; parameters are calibrated from the triaxial test results simulated with hypoplasticity and using parameters in Table 4.1

4.1.2 Results

Figure 4.3 a and b shows load-displacement curves obtained with hypoplasticity and M-C, respectively. Figure 4.4 and Figure 4.5 show the deformed meshes of the samples modelled with hypoplasticity and M-C, respectively.

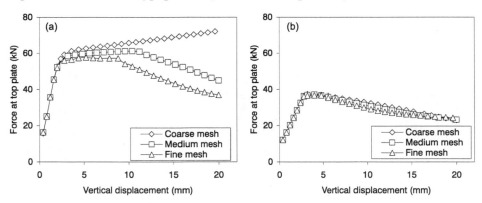

Figure 4.3: Plots of the axial load versus the displacement for biaxial test: (a) hypoplasticity; (b) Mohr-Coulomb model.

From force-displacement curves, the peak forces obtained with hypoplasticity are higher than the ones with M-C model. This behavior is expected because the M-C failure criterion considers only the major and minor principal stresses in the shear failure formulation and excludes the effect of the intermediate principle stress. In contrast, the hypoplastic constitutive equation considers all stress components. This is why the samples modelled with the M-C model give lower peak forces than the samples modelled with hypoplasticity. The force-displacement curves can be divided

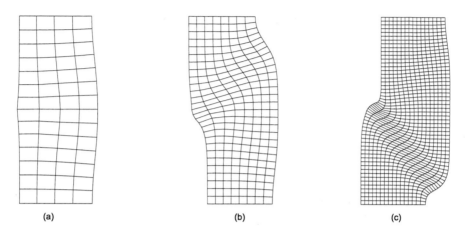

Figure 4.4: Deformed meshes at $u = 20$ mm; Biaxial tests simulation with hypoplasticity; (a) coarse mesh; (b) medium mesh; (c) fine mesh.

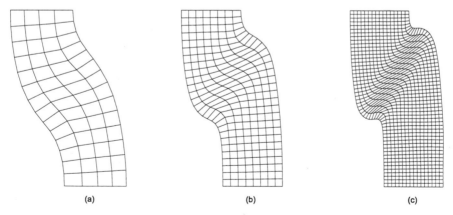

Figure 4.5: Deformed meshes at $u = 20$ mm; Biaxial tests simulated with M-C model; (a) coarse mesh; (b) medium mesh; (c) fine mesh.

into two parts. First, the part before the peak load is reached; second, after the peak force where the sample exhibits softening. However, if no shear band is formed in the sample (Fig 4.4 (a)), the force-displacement curve shows hardening with no limit. The deformed meshes shows that shear bands appear in all samples except the coarse mesh with hypoplasticity. The shear band formation in fine mesh obtained with hypoplasticity is similar to the experimental result on dense sand by Vardoulakis (48) (as shown in Figure 4.6).

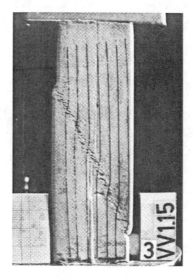

Figure 4.6: Shear band observed in a dense sand sample (48)

The shear bands pattern are different for different mesh density. Obviously, the shear band formation depends on the coarseness of the mesh. The shear band in the coarse mesh is not clearly visible; however, it is well defined in the fine mesh.

4.2 Simulation of a simple shear test

4.2.1 Simulation procedure

The simulation of an ideal plane-strain simple shear test is done by assuming a rectangular soil sample of 150 mm × 50 mm. Figure 4.7 shows the geometry and the boundaries conditions of the test. The sample is confined laterally with two rigid plates. Each plate has a hinge at the bottom which allows rotation of the plate at the base of the sample. A constant stress boundary of 100 kN/m^2 is applied to the top of the sample. The bottom of the sample is fixed in both vertical and horizontal directions. The top of the sample is free to move in vertical direction to allow dilation or contraction of the sample. The shearing process is simulated by applying velocities

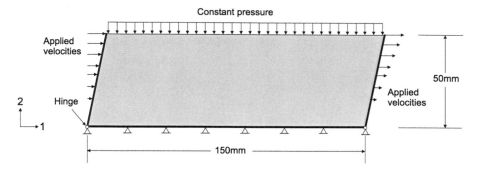

Figure 4.7: Geometry and boundaries conditions for a simple shear test

which varied from the zero at the base of the sample to the maximum at the top of the sample. The numerical stability depends on the applied velocity as described in 3.3. Therefore, the applied velocities are incrementally applied from the beginning of the simulation to 6000 step. After the calculation step reaches 6000 steps, a constant velocity of 0.9×10^{-7} is applied throughout the rest of the simulation (Fig. 4.8). The detail observations of the influence of incrementally applied velocity on the numerical stability are presented in 4.4.

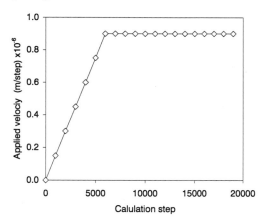

Figure 4.8: Plot of the applied velocity versus calculation step

The soil is discretized into a mesh which consists of 1875 grid zones as shown in Figure 4.9. The hypoplasticity by Wu is used to model the soil. The parameters for the hypoplasticity are taken from Kolymbas (19), as shown in Table 4.3.

C_1	C_2	C_3	C_4
-106.5	-801.5	-797.1	1077.7

Table 4.3: Hypoplastic parameters for the simulation simple shear test (19)

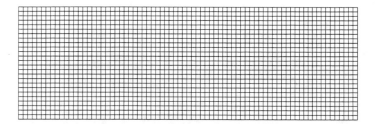

Figure 4.9: The mesh for simple shear test.

4.2.2 Simulation results

Figure 4.10 (a) and (b) show the plot of shear force at the top of the sample versus shear strain γ and the plot of vertical displacement at the middle-top of the sample versus shear strain γ, respectively. Figure 4.11 shows the deformed mesh after the sample has been sheared to $\gamma = 30\%$. The plot of the volumetric strain contours are shown in Fig. 4.12.

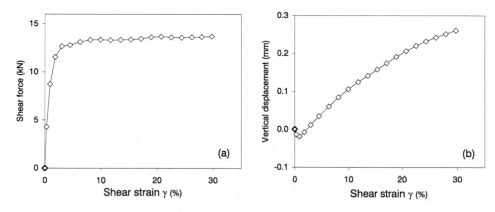

Figure 4.10: (a) Shear force at the top of the sample versus shear strain; (b) vertical displacement at middle-top of the sample versus shear strain.

From the deformed mesh, small curvatures appear at both top-corners of the sample.

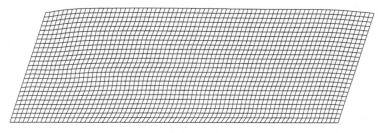

Figure 4.11: Deformed grid after the sample has been sheared; $\gamma = 30\%$

However, at the point about 10% of the sample's width away from the side boundary,

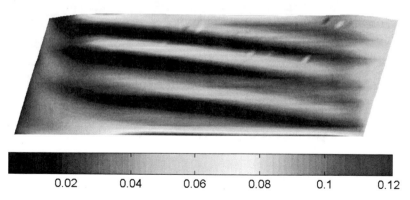

Figure 4.12: Plot of volumetric strain contour (the light color stripes are the zones which have higher volumetric strain than the surrounding zone)

these curvatures are not appeared. This due to the friction between the soil and the side plates. The plot of the volumetric strain contour shows the formation of shear bands in the soil sample. The shear bands are parallel to each other in a certain distance as also found experimentally by Tejchman (44). The pattern of shear bands formation detected experimentally by radiograph (2) is shown in Figure 4.13 for a comparison. The shear bands in simple shear experiment on sand also found by Ravuzhenko (41) as shown in Figure 4.14. The volumetric strains in the shear bands

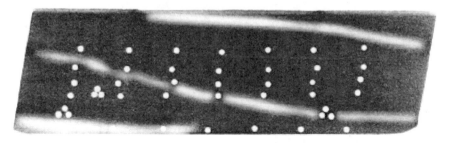

Figure 4.13: Radiographs of shear bands in a dense fine sand (2) (the white circles are lead shot used to trace the internal displacement; white lines are shear bands).

are higher than the zone outside the shear bands. The maximum volumetric strain in the shear bands is about 8% where the volumetric strain outside the shear bands is about 2%. The plot of shear force versus shear strain shows some non-smooth of the curve. This may be because of the shear bands formation in the sample.

4.3   Simulation of a spread footing on cohesionless soil

The numerical simulations of a spread footing on cohesionless soil has been done by several authors. Most simulations were done with finite element methods combined

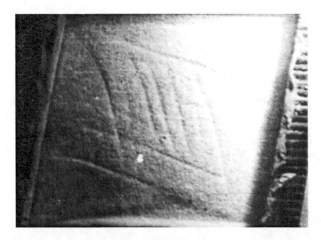

Figure 4.14: Shear bands pattern obtained by simple shear test (41)

with an elastic-perfectly plastic Mohr-Coulomb model (56; 8; 5; 7; 31; 52; 14). The simulations were aimed to determine the bearing capacity factor N_γ for spread footings. In this research, FLAC is used to calculate bearing capacity of a spread footing with hypoplasticty. The simulated results are compared with the results obtained with an elasto-plastic model and an analytical solution.

4.3.1 Simulation procedure

The geometry of the problem is shown in Figure 4.15. The soil is discretized to quadrilateral zones which are internally subdivided into two overlaid constant-strain triangular elements, as described in section 2.3. The mesh arrangement is adopted from Herle et al. (14), both sides of the vertical boundaries are horizontally fixed and the bottom boundary is fixed in both direction. A coarse mesh and a fine mesh are used to study the influence of the mesh density as shown in Figure 4.16. The rigid footing with a frictionless interface is assumed by letting the soil move freely in horizontal direction. The footing is loaded by applying a constant velocity to simulate rigid movement of the footing.

4.3.2 Selection of parameters

Soil is modelled with hypoplasticity by Wu and elastic-perfectly plastic model with the Mohr-Coulomb (M-C) yield criterion. The elastic constants for M-C model taken from Woodward et al. (52). The friction angle is varied from $25°$ to $40°$ for the parametric study of bearing capacity factors. The parameters for hypoplasticity are

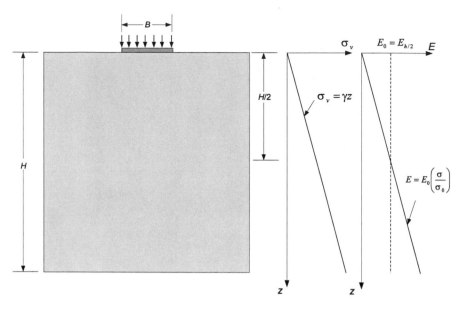

Figure 4.15: The spread footing geometry with the vertical stress and Young's modulus profiles

obtained by calibrating the model by using M-C parameters i.e. Young's modulus E, the dilatancy angle ψ and the friction angle ϕ as shown in Figure 4.17. The maximum deviatoric stress used in the calibration of hypoplasticity is

$$(\sigma_1 - \sigma_3)_{max} = \left(\frac{2 \sin \varphi}{1 - \sin \varphi} \right) \sigma_3, \tag{4.1}$$

and the slope β_A is calculated by using the definition of dilatancy angle by Vermeer et al. (50) as follows

$$\beta_B = \frac{2 \sin \psi}{1 - \sin \psi}. \tag{4.2}$$

For numerical simulation with FLAC and hypoplasticity, the dilation angle ψ used in Equation 4.2 must be carefully selected because it has some effect on the numerical stability. Figure 4.18 shows load-displacement curves of a spread footing obtained with hypoplasticity. The friction angle of $35°$ is used in all simulations and the dilatancy angle is varied. It can be seen that for $\psi = 0°$ the solution becomes unstable. The limit state cannot be obtained and the bearing capacity factor N_γ is higher than the analytical solution of Mayerhof about 136%. The solution is more stable when the different between ψ and φ is smaller. The similar problem was reported by de Borst et al. (5). Figure 4.19 shows the load-displacement curves for a smooth strip footing obtained with FEM. It can be seen that if $\psi = 0°$ the numerical instability occurred, which is similar to FLAC. The authors (5) mentioned about Figure 4.19 that "severe numerical instabilities were met, and a truly converged

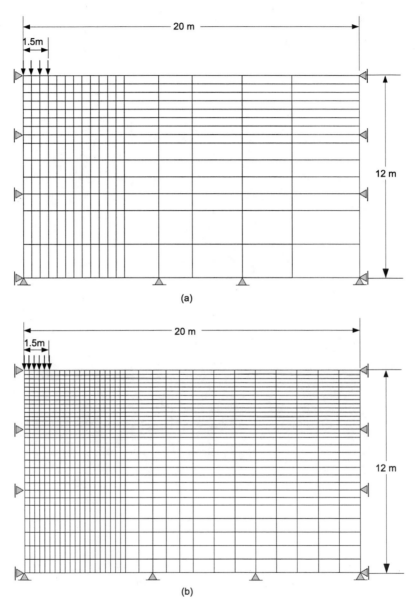

Figure 4.16: Meshes for spread footings: dimensions of the mesh are taken from Herle et al. (14) (a) Coarse mesh; (b) Fine mesh

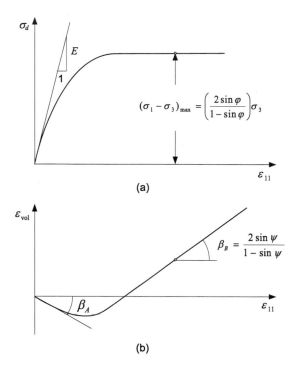

(a)

(b)

Figure 4.17: M-C parameters used in the calibration of hypoplasticity

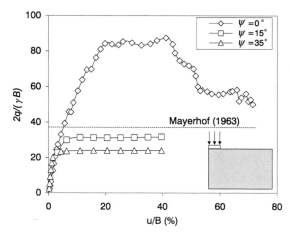

Figure 4.18: Load displacement curves for a smooth strip footing obtained with FLAC and hypoplasticity; $\varphi = 35°$ and various ψ are used in the calibration

solution could not be obtained for various stages of the loading process". Therefore, to avoid the numerical instability in FLAC, the dilatancy angle of 15° (50) is assumed in the calibration of hypoplasticity (Table 4.4).

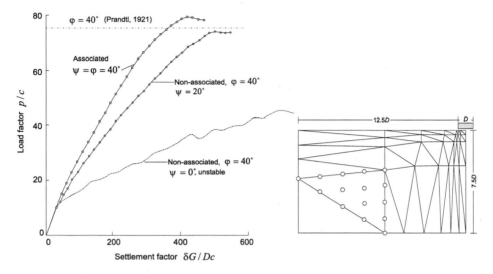

Figure 4.19: Finite element mesh and load-displacement curves for a smooth strip footing on a weightless soil. The friction angle and the cohesion are kept constant, but three different dilatancy angles are used (5).

dense sand	15°
loose sand	< 10°
normally consolidated clay	0°

Table 4.4: Typical values for dilation angles; Vermeer et al. (50)

It should be noted that, for hypoplasticity, the numerical stability also depends on the initial slope of the volumetric strain-strain curve β_A used in the calibration. The influence of β_A is shown in Figure 4.20.

It can be seen from the curves that the limit load ratio $N_\gamma = 2q/(\gamma B)$ from simulation is about 140% higher than the analytical solution of Mayerhof if $\beta_A = 0.8$. The N_γ reduces as the β_A increases. N_γ is less than the Mayerhof solution if the β_A is greater than 1.0. In addition, the stable solution cannot be obtained if β_A is less than 0.6. Therefore, the β_A must be carefully selected to avoid numerical instability. In this study, the β_A value of 1.2 is assumed in all calibrations to ensure the numerical stability. On the other hand, in case of unloading problem, such as the trapdoor problem (Fig. 4.21) or tunnels excavation (Fig 4.22), the dilatancy angle does not has strong influence on the numerical stability. The detail simulations of the trapdoor problem and tunnel excavation are presented in Section 4.4 and 4.5, respectively.

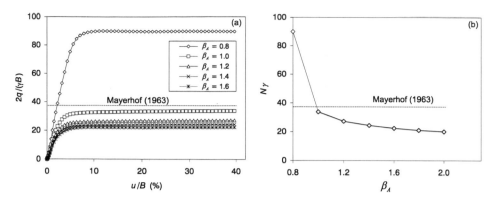

Figure 4.20: Influence of β_A used in calibration of hypoplasticity on bearing capacity factor N_γ.

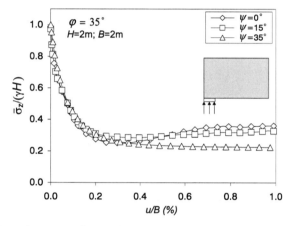

Figure 4.21: Ground reaction curves for trapdoor problem obtained with hypoplasticity show no instability when $\psi = 0°$ is used in the calibration

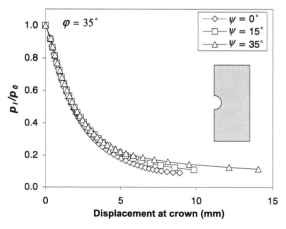

Figure 4.22: Ground reaction curves for an unlined tunnel problem obtained with hypoplasticity show no instability when $\psi = 0°$ is used in the calibration

The parameters for the parametric study of bearing capacity factors are summarized in Table 4.5 and 4.6. The influence of Young's modulus varying with depth for

E (MN/m^2)	100
ν	0.3
$\varphi(°)$	25, 30, 35, 40
$\psi(°)$	15

Table 4.5: Parameters for elastic-perfectly plastic Mohr-Coulomb model; the elastic parameters E and ν are taken from Woodward et al. (52); the friction angle is varied for parametric study; the dilatancy angle is taken from Vermeer et al. (50) and $K_0 = 1 - \sin \varphi$ is assumed.

$\varphi(°)$	C_1	C_2	C_3	C_4
25	-308.64	-5208.09	-6463.17	30464.02
30	-308.64	-3001.8	-3841.71	12452.41
35	-308.64	-2154.48	-2834.94	6863.78
40	-308.64	-1717.87	-2316.17	4468.66

Table 4.6: Parameters for hypoplasticity: the parameters are calibrated by using elastic-perfectly plastic parameters in Table 4.5.

M-C model is also concerned. Due to gravity, the soil stresses increase linearly with depth; therefore, the soil stiffness is also increasing with depth. The relationship between the stiffness of the soil and the stress as proposed by Terzaghi (12) is

$$E = E_0 \left(\frac{\sigma}{\sigma_0} \right), \qquad (4.3)$$

where E_0 corresponds to E at stress σ_0. In the simulation, the values of E_0 and σ_0 are taken at the mid-height of the model, i.e. we assume that Young's modulus is zero at the top of the model and increases linearly with depth (Fig. 4.15).

4.3.3 Simulation results

The influence of the mesh density on the simulation results is shown in Figure 4.23. Peak loads obtained with coarse mesh are lower than peak loads obtained with a fine mesh for both hypoplasticity and M-C model. Considering the influence of the soil models, the M-C model gives a higher peak load than the hypoplastic model; in addition, the M-C model with constant Young's modulus gives higher peak load than that with variable Young's modulus. The stiffness of load-displacement curve of the M-C model with varying E with depth is lower than that obtained by using the M-C model with constant E.

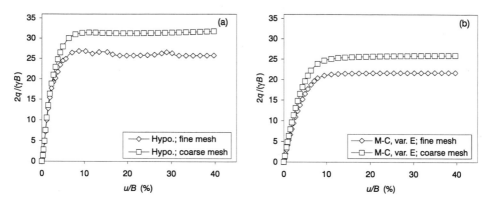

Figure 4.23: Influence of mesh density on load-displacement of footing; (a) hypoplasticity; (b) Mohr-Coulomb

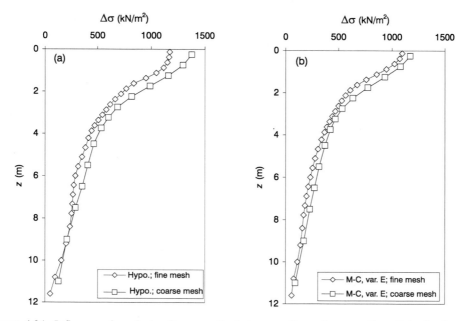

Figure 4.24: Influence of mesh density on vertical stress profile at the centerline of the footing; (a) hypoplasticity; (b) Mohr-Coulomb

The vertical stress profiles under the centerline of the footing are plotted in Figure 4.24. The vertical stress profile obtained with coarse mesh is higher than that obtained with fine mesh for the hypoplasticity, M-C with constant E and M-C with variable E. Figure 4.25 shows the plots between the normalized bearing stress $2q/(\gamma B)$ versus the normalized displacement u/B obtained from the parametric study.

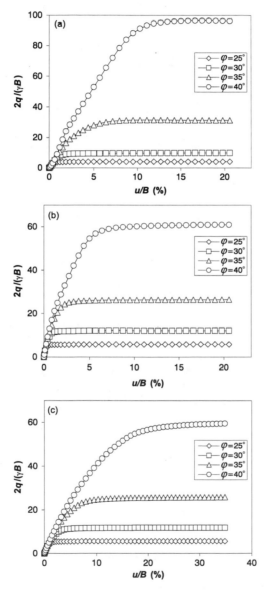

Figure 4.25: Load-displacement curves of a spread footing; (a) hypoplasticity; (b) Mohr-Coulomb with a constant Young's modulus; (c) Mohr-Coulomb with variable Young's modulus

The results show that the load-displacement curve obtained with M-C with a constant E is stiffer than that obtained with variable E. Both models reach the same limit state. This implies that the variable E affects the displacement of the footing but it does not affect the limit load. In case of hypoplasticity, the stiffness of the load-displacement curve is in between M-C with a constant E and M-C with variable E for $\varphi = 25°, 30°$ and $35°$. For $\varphi = 40°$, the stiffness of the load-displacement curve is similar to that obtained with M-C with constant E.

The bearing capacity factors N_γ are calculated from the limit stresses obtained from the load-displacement curves in Figure 4.25 by

$$N_\gamma = \left(\frac{2q}{\gamma B}\right)_{\text{limit}} \tag{4.4}$$

The N_γ values obtained with FLAC are compared with that obtained with analytical solution of Mayerhof (2). The Mayerhof's bearing capacity equation reads

$$N_\gamma = (N_q - 1)\tan(1.4\varphi'), \tag{4.5}$$

where

$$N_q = \exp^{\pi \tan \varphi'} \tan^2(45° + \frac{\varphi'}{2}). \tag{4.6}$$

Figure 4.26 shows the plots of the N_γ values obtained with FLAC and obtained with Equation 4.5 versus the friction angles.

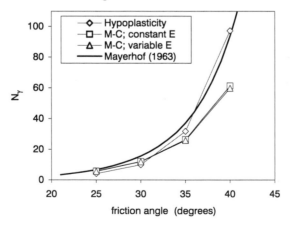

Figure 4.26: Comparison of N_γ obtained from FLAC and the analytical solution of Mayerhof

Form the plots, it can be seen that the N_γ obtained with hypoplasticity are close to that obtained with the analytical solution of Mayerhof. M-C model with constant E and variable E give the same values of N_γ. The N_γ obtained with M-C model are lower than that obtained with the analytical solution of Mayerhof. The discrepancy of the N_γ form M-C model and from the analytical solution increases as the friction angle increases. To sum up, hypoplasticity can be used to predict the bearing capacity factor with a promising results. The N_γ obtained with hypoplasticity are closer to the analytical solution than that obtained with M-C model.

4.4 Simulation of a trapdoor problem

The physical experiments of the trapdoor with plane-strain condition were done by Terzaghi (47), Vardoulakis (49) and Papamichos (38). The "ground reaction curves" and stress distributions were reported by these authors. The FEM simulations of the trapdoor were performed by Koutsabeloulis et al. (22) for a rectangular trapdoor with a plane-strain condition. The relationship between limit load ratio and various height to trapdoor width was presented. The trapdoor was also simulated with an axisymmetric finite element program by Papamichos et al. (38). The authors reported that the numerical simulation reproduced qualitatively the experimental results. The simulation of a trapdoor was done with FLAC by Kolymbas et al. (21). The elastoplastic M-C model was used in the simulation. The authors reported that the results obtained depend on the coarseness of the mesh. In this study, a trapdoor problem is simulated by using FLAC. Various ratios of overburden height and trapdoor width H/B are assumed for the parametric study.

4.4.1 Simulation procedure

The geometry of the trapdoor problem is shown in Figure 4.27. The left boundary is the plane of symmetry and free to move in vertical direction. The other boundaries are fixed in both vertical and horizontal directions.

Downward moving of the trapdoor is simulated by applying a slow constant velocity at the bottom of the mesh where the trapdoor is located. The influence of loading speed on the simulation with hypoplastic material is also studied. For a preliminary simulation with geometrical ratio $H/B=2$ and constant applied velocity of 1×10^{-6} m/step, the inertia effects have a significant influence on the solution. Figure 4.29 shows the plots of the normalized average pressure on the trapdoor versus the normalized trapdoor displacement for three different friction angles. The results reveal an influence of the applied velocity on the model behavior. The normalized average pressures on the trapdoor abruptly drop from 1.0 to 0.7 at the initial stage of load application. Afterwards, the normalized average pressures converge to constant values.

To solve the problem of the abrupt drop of the pressure on the trapdoor at the beginning of the loading, a constant applied velocity is replaced by gradually increasing

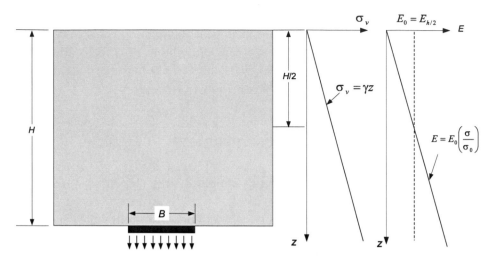

Figure 4.27: The trapdoor geometry with the definition of in-situ stress and Young's modulus

velocity as shown in Figure 4.30. After the gradually increasing scheme is applied, the abrupt change of the pressure on the trapdoor is removed as shown in Figure 4.31. The soil is modelled with hypoplasticity and M-C model. For M-C model, the elastic parameters are taken from Woodward et al. (52). The friction angle is varied from 30° to 40° for the parametric study. The dilatancy angle $\psi = 15°$ is assumed (as described in section 4.3). The parameters for hypoplasticity are calibrated on the basis of the M-C parameters. The parameters of the M-C model and hypoplasticity are summarized in Table 4.7 and 4.8. Geostatic stresses (with lateral earth pressure coefficient at rest $K_0 = 1.0$) are assumed as initial stress state of the soil. In case of the Mohr-Coulomb model, the effect of variability of Young's modulus with depth is also allowed for. The Young's modulus is varied with depth by using equation 4.3.

E (MN/m^2)	100
ν	0.3
$\varphi(°)$	30, 35, 40
$\psi(°)$	15

Table 4.7: Parameters for elastic-perfectly plastic Mohr-Coulomb model; the elastic parameters E and ν are taken from Woodward et al. (52); the friction angle is varied for parametric study; the dilatancy angle is taken from Vermeer et al. (50) and $K_0 = 1.0$ is assumed.

4.4.2   Results and discussion

Figures 4.32, 4.33, and 4.34 show the plots of normalized average pressure $\bar{\sigma}_z/(\gamma H)$ on the trapdoors versus normalized trapdoor displacements u/B for $H/B = 1$, $H/B = 2$, and $H/B = 4$. The results show that the M-C model with variable

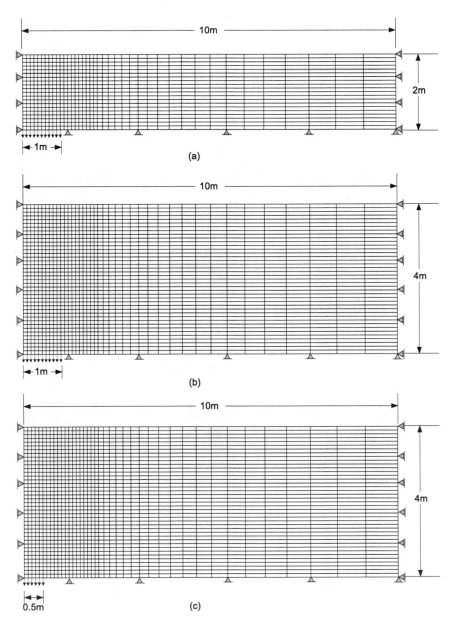

Figure 4.28: Meshes for the trapdoor problems: (a) $H/B = 1$; (b) $H/B = 2$; and (c) $H/B = 4$

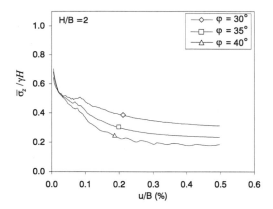

Figure 4.29: The inertial effect on load-displacement curve; hypoplastic model; $H/B = 2$

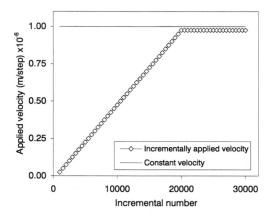

Figure 4.30: Characteristic of the incrementally applied velocity and a constant applied velocity on the trapdoor

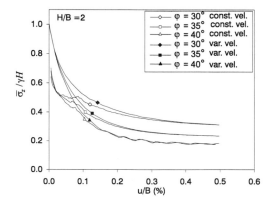

Figure 4.31: The initial effect is removed after incremental velocity scheme is applied

$\varphi(°)$	C_1	C_2	C_3	C_4
30	-308.64	-3001.8	-3841.71	12452.41
35	-308.64	-2154.48	-2834.94	6863.78
40	-308.64	-1717.87	-2316.17	4468.66

Table 4.8: Parameters for hypoplasticity: the parameters are calibrated from elastic-perfectly plastic parameters in Table 4.7.

Young's modulus E give similar curves as the M-C model with constant E. In case of $H/B = 1$, the hypoplasticity gives higher limit stress ratio than the M-C model for all values of friction angle. For $H/B = 2$, the hypoplasticity also gives higher limit stress ratio than the M-C model for $\varphi = 30°$ and $35°$. However, both hypoplasticity and the M-C model give almost the same limit stress ratio for $\varphi = 40°$. For $H/B = 4$, hypoplasticity and the M-C model give the same limit stress ratio for all values of φ.

The average vertical stress $\bar{\sigma}_z$ can be calculated by using an analytical solution for silos by Janssen (20). The Janssen's equation for cohesionless soil reads

$$\bar{\sigma}_z = \frac{\gamma}{K_0 m \tan \varphi} \left(1 - \exp^{(-K_0 m \tan \varphi)z} \right), \tag{4.7}$$

where

$$m = \frac{\text{Perimeter of trapdoor}}{\text{Area of trapdoor}}. \tag{4.8}$$

In case of a rectangular trap door with plane strain conditions where the out-of-plane length goes to infinity, the m value is calculated from

$$m = \lim_{L \to \infty} \frac{2(B + L)}{BL} = \frac{2}{B}. \tag{4.9}$$

Equation 4.9 is substituted into Equation 4.7. Thus we finally have

$$\bar{\sigma}_z = \frac{\gamma B}{2K_0 \tan \varphi} \left(1 - \exp^{(-2K_0 \tan \varphi/B)z} \right). \tag{4.10}$$

By giving the H/B ratio, the average vertical stress $\bar{\sigma}_z$ profiles over the trapdoor can be plotted. The $\bar{\sigma}_z$ at the trapdoor level is calculated by setting $z = H$. This analytical solution is compared with the results from FLAC.

Figures 4.35 to 4.37 show the plots of the average vertical stress profile along the centerline of the trapdoor. The results are compared with the analytical solution of Janssen (Eq. 4.10). From the results, it can be seen that the stress profiles obtained with FLAC agree well with the analytical solution for $H/B = 1$. The average vertical stresses on the trapdoor ($z = 2$ m) obtained with hypoplasticity are similar to the analytical solutions. In contrast, the M-C model underestimates the average stress

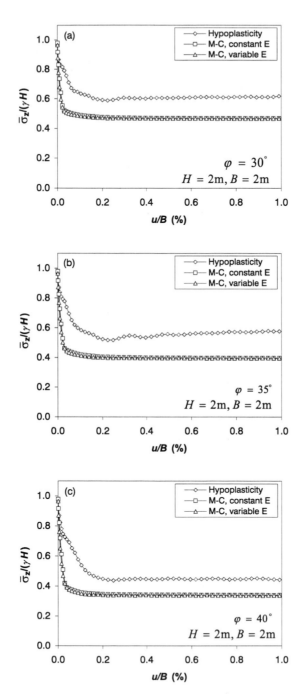

Figure 4.32: Plots of the normalized average pressures on the trapdoors versus the normalized trapdoor displacements for $H = 2$ m, $B = 2$ m; (a) $\varphi = 30°$; (b)=$\varphi = 35°$; (c) $\varphi = 40°$

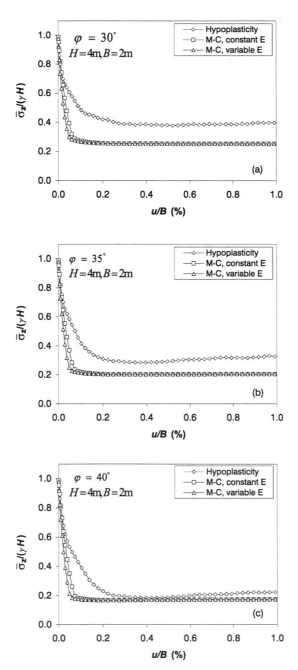

Figure 4.33: Plots of the normalized average pressures on the trapdoors versus the normalized trapdoor displacements for $H = 4$ m, $B = 2$ m; (a) $\varphi = 30°$; (b)=$\varphi = 35°$; (c) $\varphi = 40°$

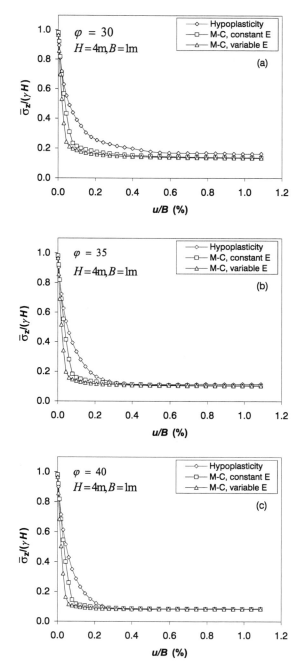

Figure 4.34: Plots of the normalized average pressures on the trapdoors versus the normalized trapdoor displacements for $H = 4$ m, $B = 1$ m; (a) $\varphi = 30°$; (b)=$\varphi = 35°$; (c) $\varphi = 40°$

on the trapdoor for all values of φ. For $H/B = 2$, the shapes of the stress profiles obtained from FLAC differ from the ones obtained with analytical solution for both hypoplasticity and M-C model. Hypoplasticity overpredicts the stress profile and M-C model underpredicts the stress profile. However, the stresses at the trapdoor level agree well with the analytical solutions in case of hypoplasticity for $\varphi = 30°$ and $\varphi = 35°$. For $H/B = 4$, the stress profiles are totally different from the analytical ones. However, the stress profiles obtained from FLAC are similar to the stress profile experimentally obtained by Terzaghi (47), as shown in Figure 4.38. The experiment was done on a compacted sand with $H/B = 4.3$. From the results, they imply that for $H/B = 1$, the failure mechanism obtained with FLAC is similar to the failure mechanism assumed for the Janssen's equation. For $H/B = 2$ and $H/B = 4$, the failure mechanisms obtained with FLAC differ from the Janssen's failure mechanism.

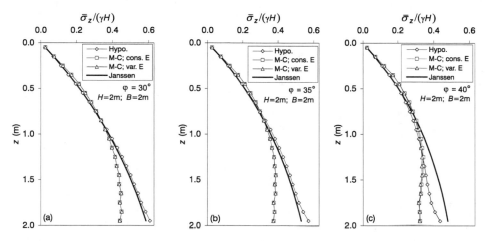

Figure 4.35: Average vertical stress profiles along the centerline of the trapdoor for $H = 2$ m, $B = 2$ m; (a) $\varphi = 30°$; (b) $\varphi = 35°$; (c) $\varphi = 40°$

Equation 4.10 is used to calculate the limit stress ratios at the trapdoor level ($z = H$) for various values of H/B. Figure 4.39 shows the plots of the limit stress ratios obtained with FLAC and the limit stress ratios obtained with the analytical solution (Eq. 4.10). For $H/B = 1$ and $H/B = 2$, the results obtained with hypoplasticity are close to the analytical solutions for $\varphi = 30°$ and $\varphi = 35°$. For $H/B = 4$, hypoplasticity underpredicts the limit stress ratio. The underprediction is expected because the failure mechanism of the soil is differ from that assumed in the analytical solution. In case of M-C model, it underpredicts the limit stress ratio for all H/B and all φ.

To summarize, hypoplasticity by Wu implemented in FLAC can be used to simulate the trapdoor problem with promising results. The stress profiles obtained with hypoplasticity are close to the analytical solution for $H/B = 1$. In case of $H/B = 4$,

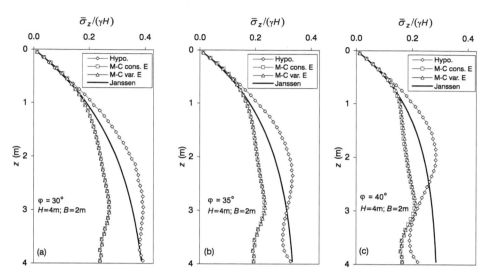

Figure 4.36: Average vertical stress profiles along the centerline of the trapdoor for $H = 4$ m, $B = 2$ m; (a) $\varphi = 30°$; (b) $\varphi = 35°$; (c) $\varphi = 40°$

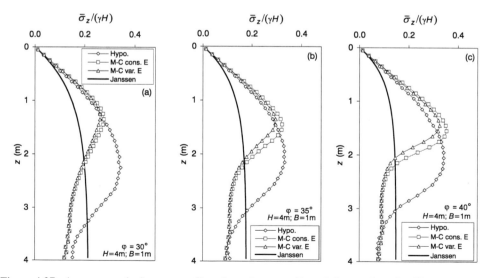

Figure 4.37: Average vertical stress profiles along the centerline of the trapdoor for $H = 4$ m, $B = 1$ m; (a) $\varphi = 30°$; (b) $\varphi = 35°$; (c) $\varphi = 40°$

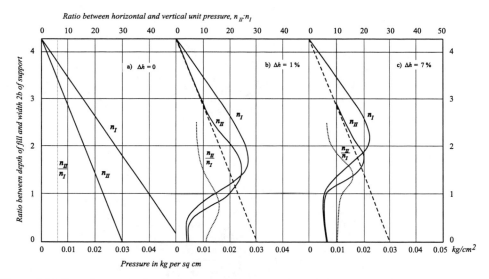

Figure 4.38: Experimentally obtained stress profile over a plane (47); the sand is in a compacted state; $H/B = 4.3$; n_I is vertical stress; n_{II} is horizontal stress; (a) for the initial state; (b) for trapdoor movement $\Delta h = 1\%$; (c) for trapdoor movement $\Delta h = 7\%$

hypoplasticity reproduces qualitatively the stress profile from the experimental result.

4.5 Simulation of an unlined circular tunnel

For a tunnel in geostatic primary stress, the analytical solution for stresses and displacement around the tunnel is extremely complicated (33). Therefore, FLAC is used to simulate an unlined circular tunnel to obtain "ground reaction curves" and the stress distribution around the tunnel.

4.5.1 Simulation procedure

Figure 4.40 shows the geometry and boundary conditions of the problem. The tunnel with diameter of 2 m is located 20 m below the ground surface. The soil is discretized into quadrilateral zones (as described in section 2.3). Two types of meshes (Fig. 4.41 are used to investigate the influence of mesh arrangement on the numerical simulations. Both vertical boundaries of the mesh are fixed horizontally, the bottom boundary is fixed in both directions and the top boundary is free. Geostatic primary stresses are assumed in the simulation with $K = 1.0$. The soil is modelled with the M-C model and hypoplasticity. In case of the M-C model, the variation of Young's modulus with depth (Eq. 4.3) due to the dependency of the stiffness on pressure is

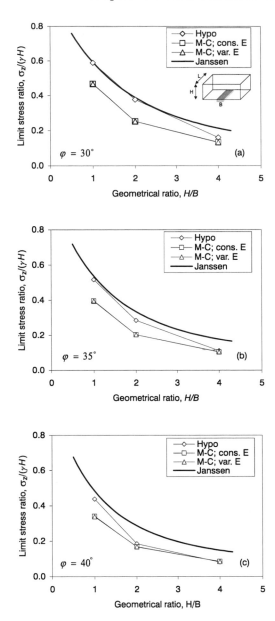

Figure 4.39: Comparison of numerical simulation with Janssen's equation: plot of limit stress ratio versus geometrical ratio; (a) $\varphi = 30°$; (b) $\varphi = 35°$; (c)$\varphi = 40°$

taken into account in the simulations. The parameters for the M-C model are shown in Table 4.9 and the parameters for hypoplasticity are shown in Table 4.10.

The excavation of the tunnel is simulated by using the load reduction method (43) and (55). Figure 4.42 shows the schematic diagram of load reduction method. First, the elements enclosed within the excavation boundary are removed, and the nodal forces N_0 along the excavation boundary are calculated from initial state of stress

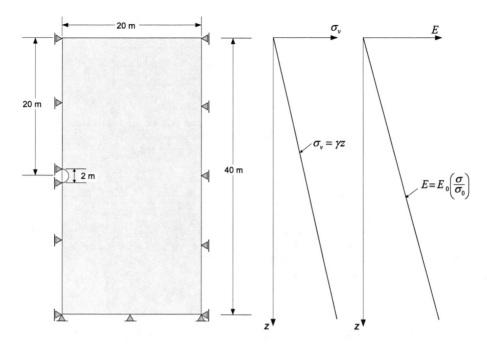

Figure 4.40: Geometry of a circular tunnel in cohesionless soil

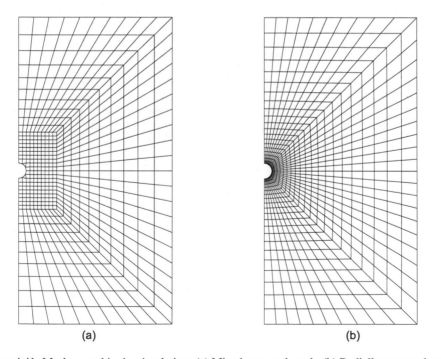

(a)   (b)

Figure 4.41: Meshes used in the simulation: (a) Mixed-arranged mesh; (b) Radially-arranged mesh

parameter	variable Young's modulus
$E(\text{MN/m}^2)$	$5z(\text{m})$
ν	0.3
$\varphi(°)$	25
$\psi(°)$	25
$c(\text{kN/m}^2)$	0

Table 4.9: Mohr-Coulomb parameters for the circular tunnel simulation; the elastic parameters are taken from Woodward et al. (52); the friction angle $\varphi = 25°$ is assumed.

C_1	C_2	C_3	C_4
-64.1	-5375.6	-1820.1	10933.7

Table 4.10: Hypoplastic parameters for the circular tunnel simulation; the parameters are calibrated by using elastic-perfectly plastic parameters in Table 4.9.

σ_0. These calculated nodal forces N_0 are applied back to the tunnel boundary to provide equilibrium. The applied nodal forces around the tunnel are then gradually reduced to $(1 - \beta)N$. The load factor is in the range of $0 < \beta < 1$. In this research, the nodal forces are reduced until they become zero or until the model reaches the limit state (i.e. large deformation occur within one load increment).

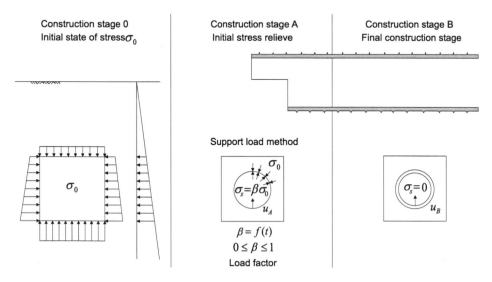

Figure 4.42: Two-dimensional simulation of the three-dimensional effect (43)

4.5.2 Simulation results

Figure 4.43 shows the plots of the vertical stress σ_z versus the displacement at the crown of the tunnel of mesh type I and type II. In this study, these plots are called "ground reaction curves". The curves obtained with hypoplasticity reach the limit of 0.26 for mesh type I and 0.23 for mesh type II (Fig. 4.43 a). In case of M-C model, the curves reach the limit value of 0.21 for mesh type I and 0.17 for mesh type II (Fig. 4.43 b). The shape of ground reaction curve obtained with hypoplasticity shows no separation between elastic and elastic part. In contrast, M-C model shows linear relationship at the elastic stage followed by plastic deformation. The ground reaction curves obtained with mesh type I lies above the ones obtained with mesh type II for both hypoplasticity and M-C model.

The vertical stress profiles and horizontal stress profiles along the horizontal axis

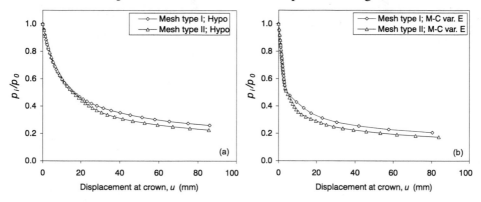

Figure 4.43: Influence of mesh type on ground reaction curves: (a) Hypoplasticity; (b) Mohr-Coulomb with varying Young's modulus

of the tunnel are shown in Figure 4.44 for the mesh type I and type II. Figure 4.45 and 4.46 show the vertical and horizontal stress profiles along the vertical axis of the tunnel for mesh type I and type II, respectively. The results shows that the vertical stress profiles along vertical axis obtained with hypoplasticity are lower than the ones obtained with M-C model for both meshes type I and type II. For horizontal stress σ_h profile along vertical axis, the results obtained with hypoplasticity are also lower than the ones obtained with M-C model for both mesh type I and type II. In addition, the plastic zones around the tunnel are clearly defined in case of M-C model. In contrast, there are no clearly defined plastic zones in case of hypoplasticity.

In summary, the arrangement of the mesh does not show significant effect in the numerical results for M-C model and hypoplasticity.

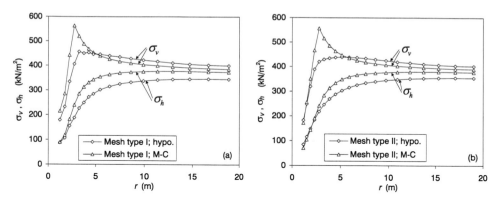

Figure 4.44: Stress profiles along the horizontal centerline of the tunnel; (a) mesh type I; (b) mesh type II

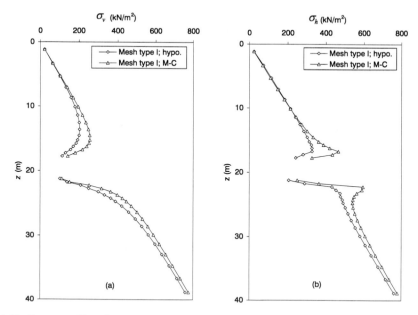

Figure 4.45: Stress profiles along the vertical centerline of the tunnel; Mesh type I; (a) vertical stress; (b) horizontal stress

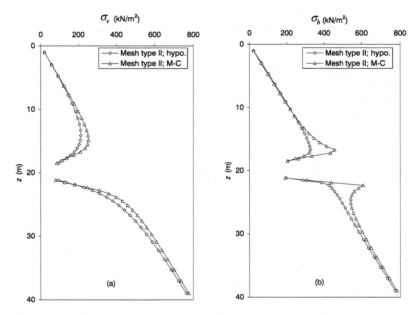

Figure 4.46: Stress profiles along the vertical centerline of the tunnel; Mesh type II; (a) vertical stress; (b) horizontal stress

Chapter 5

Hypoplasticity for normally consolidated clay

The hypoplastic constitutive equation was originally designed for sand. However, loose sand and normally consolidated clay have some common behavior. Therefore, hypoplastic constitutive equations are used to model normally consolidated (NC) Bangkok Clay. Various hypoplastic constitutive equations are examined to find the suitable constitutive equation to be used in Chapter 6. This research focusses on the behavior of cohesive soils in short term conditions; thus, the calibrations are made to obtain the best fit to undrained test results.

5.1 Behavior of normally consolidated clay

5.1.1 Laboratory experimental results

Many experimental results showed that cohesive soils allow to normalize the stress deviator. In other words, strength and stiffness are proportional to the effective pressure. Figure 5.1a shows idealized triaxial compression test results of a homogeneous clay. Two samples were tested with different consolidation stresses σ'_c, then sheared until failure. If the deviator stress is normalized by the consolidation stress, $(\sigma_1 - \sigma_3)/\sigma'_c$, the two curves coincide (Fig. 5.1 b). Ladd et al. (24), (25) pointed out that, for the same soil, the normalized curve can be used to describe the behavior of other normally consolidated samples with different consolidation stress σ'_c, when sheared in the same type of test. He also noted that pore pressure and hence effective stress paths can be also normalized with the consolidation pressures. Figure 5.2 shows the plots of shear stresses normalized with consolidation stresses versus shear strain of simple shear tests on normally consolidated Maine clay Ladd et al. (25). The curves almost coincide; a small deviation of the plots mainly comes from the sample inhomogeneity and large differences in the vertical consolidation stress. As we have seen, normalization provides a framework for comparing and relating the behavior of different cohesive soils. However, if the structure of the soil is severely damaged when loaded beyond its "apparent" maximum past pressure, the normalization is no more reasonable. For instance, tests with quick clay and cemented soil

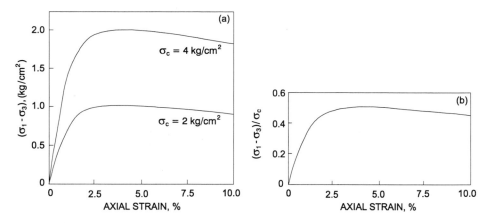

Figure 5.1: Example of normalized behavior using idealized triaxial compression test data for homogeneous clay (25)

do not allow normalization because the soil structure is significantly changed during consolidation to higher stresses.

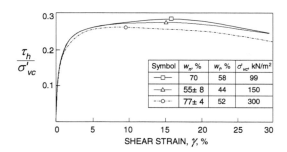

Figure 5.2: Normalized simple shear test data for normally consolidated Maine organic clay (25)

Lade et al. (27) used a true-triaxial apparatus for cubical specimens of remoulded Grundite Clay to investigate the stress-strain and pore pressure behavior under three unequal principal stresses. Figure 5.3 shows the plots of normalized deviatoric stresses $(\sigma_1 - \sigma_3)/\sigma_c'$ and normalized pore pressure changes $\Delta u/\sigma_c'$ versus the major principal strain ε_1, for three different consolidation pressures. It can be seen that the initial undrained moduli increase with increasing consolidation pressures; the pore pressures increase to the values at failure, which are almost proportional to the initial consolidation pressure.

Lade et al. (27) remarked that the three-dimensional stress-strain and strength characteristics of Grundite Clay are very similar to those observed for sand. A normally consolidated clay has only one equilibrium at a dry density that depends on the consolidation pressure possible failure surface. For a given clay, the normalized stress-strain relations show a characteristic type of behavior that does not vary much with consolidation pressure. Lade et al. (27) also concluded that the basic framework for

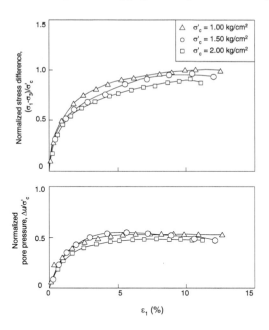

Figure 5.3: Normalization behavior of Grundite Clay Lade et al. (27)

stress-strain models developed from concepts of plasticity theory should be similar for sand and normally consolidated remoulded clay (26).

Balasubramaniam et al. (1) conducted a large number of triaxial tests on normally consolidated (NC) Bangkok clay. The undisturbed samples were taken at depths of $5.5\,\text{m} - 6\,\text{m}$. The pre-consolidation pressure of the clay is $69\,\text{kN/m}^2$. The average index properties and natural water contents are shown in Table 5.1.

natural water content	112-130%
Liquid limit	$118 \pm 2\%$
Plastic limit	$43 \pm 2\%$
Plasticity index	$75 \pm 4\%$

Table 5.1: Average index properties and natural water content of normally consolidated Bangkok Clay Balasubramaniam et. al (1)

The isotropically consolidated undrained (CIU) and isotropically consolidated drained (CID) triaxial compression tests were carried out on samples with stress control. The samples were consolidated isotropically to stresses of $138\,\text{kN/m}^2$, $207\,\text{kN/m}^2$, $276\,\text{kN/m}^2$, $345\,\text{kN/m}^2$ and $414\,\text{kN/m}^2$; afterward, the samples were sheared with increasing axial stress and under constant cell pressure. Figure 5.4 (a) shows the plots of deviatoric stress versus shear strain $\varepsilon_s = 2(\varepsilon_1 - \varepsilon_3)/3$ and Figure 5.5 (a) shows the plots of pore pressure versus shear strain of the CIU triaxial tests. The deviatoric stress and pore pressure are normalized by consolidation pressure as shown in Figure 5.4 (b) and 5.5 (b) respectively. The normalized stress-strain curves and normalized pore pressure-strain curves almost coincide. In addition, the effective stress paths can be also normalized by the consolidation stresses (Fig. 5.6). Figure 5.7a shows the plots of the peak stress condition of the test. The slope M of the critical state line in the (q, p') plot is 1.05 which corresponds to $\varphi' = 26.5°$. Figure 5.7 (b) shows the plots of the points at failure in the $(w, \log p')$ plot. In this plot, the water content at failure varies linearly with $\log p'$. The line is parallel to the isotropic consolidation line (ICL).

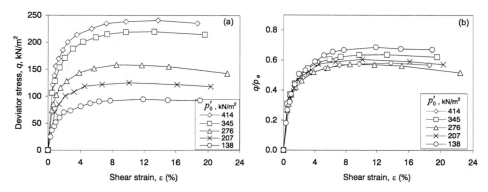

Figure 5.4: CIU triaxial tests on NC Bangkok Clay (1); (a) plots of deviatoric stress versus strain; (b) plots of normalized deviatoric stress versus strain

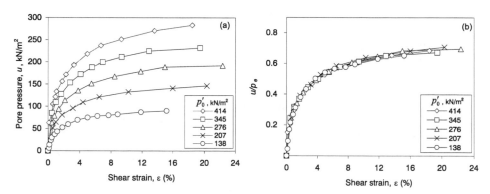

Figure 5.5: CIU triaxial tests on NC Bangkok Clay (1); (a) plots of pore pressure versus strain; (b) plots of normalized pore pressure versus strain

The normalization was done for stress-strain curves of drained triaxial tests as well; the normalized curves almost coincide (Fig. 5.8). Figure 5.9a shows the peak stress points in (q, p') plot. The slope M of the critical state line in (q, p') plot is 0.92 which corresponds to $\varphi = 23.5°$. The value of M obtained for drained test is lower than the one for undrained test. Balasubramaniam et al. (1) and James et al. (18) suggested that, for undrained condition $p'_f/p'_0 < 1$ the peak stress envelopes correspond to the Hvorslev failure envelope. However, for the drained test $p'_f/p'_0 \geq 1$ the peak stress envelope coincides with the critical state line.

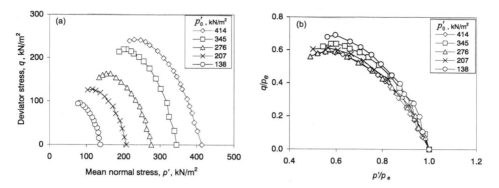

Figure 5.6: CIU triaxial tests on NC Bangkok Clay (1); (a) plots of effective stress paths; (b) plots of normalized effective stress paths

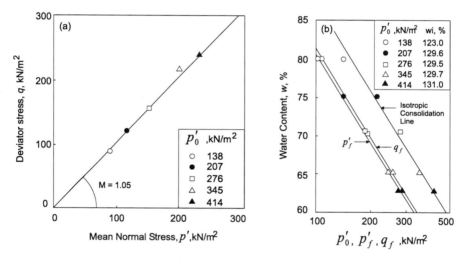

Figure 5.7: Peak stress condition of the CIU triaxial test on NC Bangkok Clay (1) (a) peak stress in (q, p') plot; (b) peak stress in $(w, \log p')$ plot

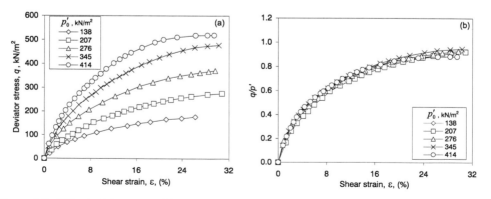

Figure 5.8: CID triaxial tests on NC Bangkok Clay (1); (a) plots of deviatoric stress versus strain; (b) plots of normalized deviatoric stress versus strain

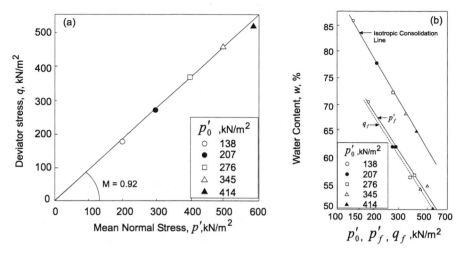

Figure 5.9: Peak stress condition of the CID triaxial test on NC Bangkok Clay (1) (a) peak stress in (q, p') plot; (b) peak stress in $(w, \log p')$ plot

Figure 5.9 (b) shows the plots of the critical states in $(w, \log p')$ plot. This line is parallel to the isotropic consolidation line and the slope of this line is 0.51. Balasubramaniam et al. (1) concluded that the normally consolidated Bangkok clay exhibits normalized behavior that is not dependent on the consolidation pressure.

5.1.2 Hypoplasticity for normally consolidated clay

The previous sections describe normalized stress-strain curves and pore-pressure-strain curves of different types of normally consolidated clay. This section describes the normalized behavior of the hypoplastic constitutive equation of Wu et al. (53). This hypoplastic constitutive equation also exhibits normalized behavior similar to the experimental data. It can be explained by the fact that the relation $\overset{\circ}{\sigma} = \mathbf{H}(\sigma, \dot{\varepsilon})$ is homogeneous in σ, i.e.

$$\mathbf{H}(\lambda\sigma, \dot{\varepsilon}) = \lambda^n \mathbf{H}(\sigma, \dot{\varepsilon}) \tag{5.1}$$

Considering the degree of homogeneity, knowing that $d\sigma/d\varepsilon = \dot{\sigma}/\dot{\varepsilon}$ is the stiffness, we infer that $(\dot{\sigma}/\dot{\varepsilon})\,|_{\lambda\sigma} = \lambda^n(\dot{\sigma}/\dot{\varepsilon})\,|_\sigma$. It can be said that, if the stresses is increased by a factor λ, the stiffness is also increased by the factor λ^n. If the deviator is normalized by the initial stress, the same plots are obtained as shown in Figure 5.10. $n = 1$ implies that the friction angle is independent of the stress level and all material constants are dimensionless. Therefore, hypoplastic constitutive equation can be used to model NC clay. Detailed studies are presented in the following sections.

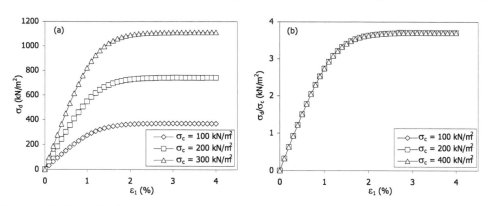

Figure 5.10: Normalization of the deviatoric stress by confining pressure; the test was simulated with hypoplasticity by Wu; parameters are taken from Kolymbas (19); (a) plots of deviatoric stress; (b) plot of normalized deviatoric stress

5.2 Calibration of hypoplasticity for Bangkok Clay

The hypoplasticity by Wu with and without structure tensor and by von Wolffersdorff are selected to model NC Bangkok Clay. The models are calibrated with conventional triaxial test results. The performances of each model are examined. Then a suitable model is selected to be used in modelling of shield tunnelling in chapter 6.

5.2.1 Hypoplasticity by Wu

Calibration with CID triaxial test results

The parameters for hypoplasticity are obtained by calibrating the model on the basis of triaxial test results of Balasubramaniam et al. (1). This study focusses on the behavior of clay around tunnels in short term conditions (undrained); therefore, a calibration to obtain a best fit to the CIU triaxial test results is applied. If the CID triaxial test results are used, a trial-and-error method is used to obtain a best fit to the CIU triaxial test results. However, if the CIU triaxial test results are used, the calibration can be done with the test results directly (as described in 5.2.1).

Four parameters are required for the hypoplastic constitutive equation, viz. C_1, C_2, C_3 and C_4; therefore, four known values from the test results, viz. E_i, β_A, β_B and $(\sigma_1 - \sigma_2)_{\max}$ (see Fig. 5.11) are used in the calibration. The calibration procedure is described in Appendix A.

The calibration was done by using the stress-strain curve at the consolidation pressure 276 kN/m^2. The calibrated parameters are shown in Table 5.2. These parameters were used to simulate triaxial compression tests with three consolidation pressures for both undrained and drained conditions. Figure 5.12a shows the plots of deviatoric stress versus axial strain curves for undrained triaxial tests and Figure

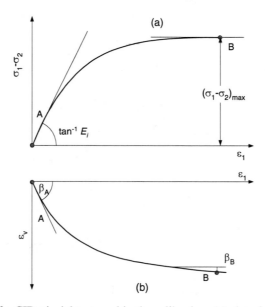

Figure 5.11: Results of a CID triaxial test used in the calibration: (a) plot of deviatoric stress versus strain, (b) plot of volumetric stress versus strain

C_1	C_2	C_3	C_4
-5.49	-61.20	-43.58	-12.73

Table 5.2: Parameter for Bangkok clay calibrated from CID triaxial test results

5.12b shows the plots of pore pressure versus axial strain. Figure 5.13a shows the plots of the undrained stress paths and Figure 5.13b shows the plots of peak stresses in $(e, \log p')$ plot. Figure5.14a shows the plot of deviatoric stress versus axial strain, and Figure 5.14b shows the plots of volumetric strain versus axial strain for the CID triaxial tests.

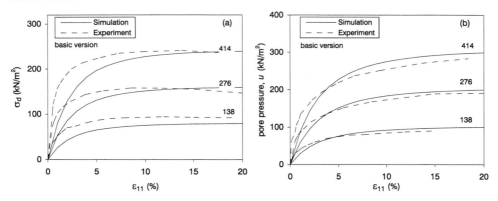

Figure 5.12: Simulations of CIU triaxial tests with hypoplasticity by Wu; the parameters in Table 5.2 are used; (a) plots of deviator stress versus strain; (b) plots of pore pressure versus strain

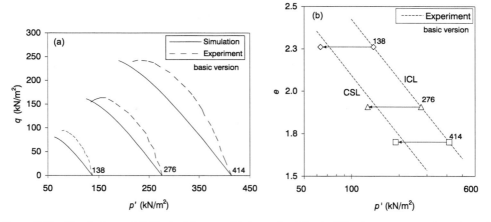

Figure 5.13: Simulations of CIU triaxial tests with hypoplasticity by Wu; the parameters in Table 5.2 are used; (a) stress paths in (q, p') plot; (b) peak stresses in $(e, \log p')$ plot

From the results, it can be seen that the stress-strain curves at the initial stage of the tests obtained with hypoplasticity are lower than the experimental stress-strain curves. However, the stresses at limit state are in good agreement with the experimental results for consolidation pressures of 276 kN/m^2 and 414 kN/m^2. In case of consolidation pressure of 138 kN/m^2, the stress at limit state from hypoplasticity is lower than the experimental result. The plots of pore pressure versus strain obtained from the simulations are slightly higher than the experimental results; still, the predictions of pore pressures at limit state are in good agreement with the experimental

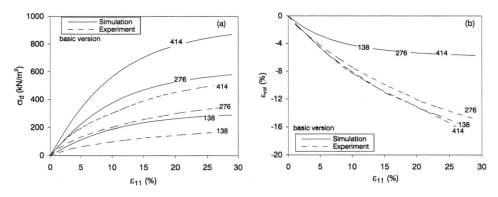

Figure 5.14: Simulations of CID triaxial tests with hypoplasticity by Wu; the parameters in Table 5.2 are used; (a) plots of deviator stress versus strain; (b) plots of volumetric strain versus strain

results. The stress paths obtained from the simulations reach the same critical state line; however, before reaching the critical state, the deviatoric stresses are lower than those from the experimental results.

For drained triaxial test, the stress-strain curves obtained with hypoplasticity are significantly higher than the experimental results. In contrast, hypoplasticity notably underpredicts the volumetric strain for all consolidation pressures.

Calibration with CIU triaxial test results

The parameters for hypoplasticity can be also calibrated by using undrained triaxial results. In this case, the plots of deviator stress versus strain and excess pore pressure versus strain, as shown in Figure 5.15, are used. For the hypoplastic constitutive equation, which has four parameters, the four known values from experimental results must be used. In case of undrained triaxial test, known values are initial tangent modulus, E_i, maximum deviatoric stress, $(\sigma_1 - \sigma_3)_{\text{max}}$, initial slope of pore pressure versus strain curve, $\Delta u_i / \Delta \varepsilon_1$, and the maximum pore pressure, u_{max}. The calibration details are described in Appendix A.

The undrained triaxial test results of NC Bangkok clay (Fig. 5.4 and Fig. 5.5) are used to calibrate hypoplasticity. The calibration was done on the sample consolidated by a pressure of 276 kN/m². Table 5.3 shows the parameters obtained from the undrained triaxial test. These parameters are used to simulate triaxial compres-

C_1	C_2	C_3	C_4
-6.44	-83.03	-51.77	-40.68

Table 5.3: Parameters for Bangkok clay calibrated from CIU triaxial test results; consolidation pressure of 276 kN/m².

sion tests for different consolidation pressures for both drained and undrained cases.

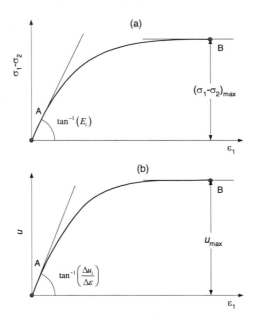

Figure 5.15: Results of an undrained triaxial test used in the calibration: (a) stress-strain curve, (b) excess pore pressure versus strain curve

Figure 5.16 (a) shows the plots of deviatoric stress versus strain; and Figure 5.16 (b) shows the plots of pore pressure versus strain for three consolidation stresses. Figure 5.17 (a) shows the effective undrained stress paths and Figure 5.17 (b) shows the peak stresses in $(e, \log p')$ plot.

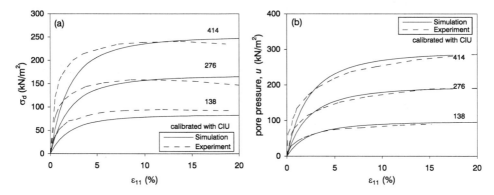

Figure 5.16: Simulations of CIU triaxial tests with hypoplasticity by Wu; the parameters in Table 5.3 are used; (a) plots of deviator stress versus strain; (b) plots of pore pressure versus strain

From the simulation results of CIU triaxial tests, the stiffnesses are lower than the experimental results. However, the stresses at limit state from the simulations are close to the experimental results except for consolidation pressure of $138 \, \text{kN/m}^2$. The plots of pore pressure versus strain from the simulations are close to the experimental results for all consolidation pressures. The end of stress paths of the simulations

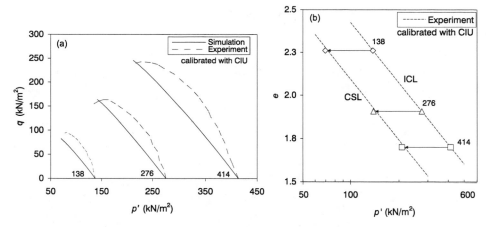

Figure 5.17: Simulations of CIU triaxial tests with hypoplasticity by Wu; the parameters in Table 5.3 are used; (a) stress paths in (q, p') plot; (b) peak stresses in $(e, \log p')$ plot

and experiment reach almost the same point on the critical state line. This can be seen in the plots of the peak stresses in $(e, \log p')$ diagram. However, before reaching the critical state, the deviatoric stresses from the simulations are lower than the experiments. The simulations are also performed for CID triaxial tests with different consolidation pressures. The results are shown in Figure 5.18.

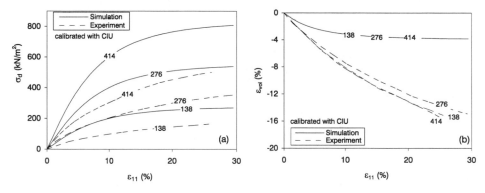

Figure 5.18: Simulation of CID triaxial tests with hypoplasticity by Wu; the parameters in Table 5.3 are used; (a) plots of deviator stress versus strain; (b) plots of volumetric strain versus strain

From the CID triaxial test results, it can be seen that the hypoplasticity by Wu overpredicts the deviator stresses for all consolidation pressures. On the other hand, the model underpredicts the volumetric strain for all consolidation pressures. However, the undrained behavior is the major concern in this research; therefore, the hypoplasticity by Wu is fairly suitable for undrained problems.

5.2.2 Hypoplasticity with structure tensor

The hypoplastic constitutive equation with structure tensor (as described in section 3.4) is used to model normally consolidated Bangkok Clay. The calibration method is similar to the basic version of hypoplasticity except that two additional parameters (μ and λ) and an initial structure tensor must be taken into consideration. The drained triaxial test results at the consolidation pressure of 276 kN/m^2 are used here. A trial-and-error method is used to obtain the best fit to the undrained triaxial test results. In addition, trial-and-error is also used to obtain the parameters μ and λ and the initial structure stress $Q1$. Table 5.4 shows the calibration results based on experimental results of Balasubramaniam et al. (1). The calibration results in Table 5.4

C_1	C_2	C_3	C_4	μ	λ	Initial isotropic structure stress
-7.69	-95.24	-68.59	-44.92	111	20	-8.0 kN/m^2

Table 5.4: Calibration results for hypoplasticity with structure tensor

are used to simulate triaxial compression tests for drained and undrained conditions with three consolidation pressures (138 kN/m^2, 276 kN/m^2 and 414 kN/m^2).
Figure 5.19 (a) shows the plots of deviatoric stress versus axial strain of undrained triaxial tests and Figure 5.19 (b) shows the plots of pore pressure versus axial strain for three consolidation stresses. Figure 5.20 shows the plots of the undrained stress paths. From the results of undrained triaxial tests, the deviatoric stresses obtained

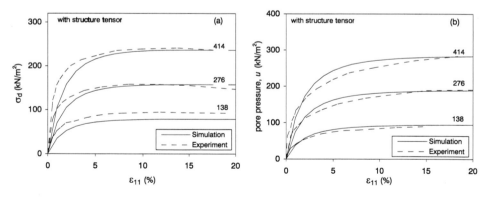

Figure 5.19: Simulations of CIU triaxial tests with hypoplasticity with structure tensor; the parameters in Table 5.4 are used; (a) plots of deviator stress versus strain; (b) plots of pore pressure versus strain

from the simulations with hypoplasticity are slightly lower than the experimental results for consolidation stresses of 138 kN/m^2 and 414 kN/m^2. However, for consolidation stress of 276 kN/m^2, the deviator stresses from simulation are in good agreement with the experimental results. The pore pressures obtained from the simulations are slightly higher than the experimental results. Moreover, hypoplasticity with structure tensor gives good predictions of the pore pressure at limit state. The

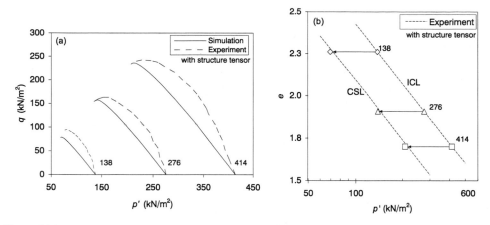

Figure 5.20: Simulations of CIU triaxial tests with hypoplasticity with structure tensor; the parameters in Table 5.4 are used; (a) stress paths in (q, p') plot; (b) peak stresses in $(e, \log p')$ plot

end of the stress paths from the simulations reach the critical state line close to the experiments. This can be seen in the $(e, \log p')$ plot also. However, before reaching the critical state line, the deviatoric stresses from simulations are lower than the experiments.

The CID triaxial tests are also simulated with hypoplasticity with structure tensor. Figure 5.21 (a) shows the plots of the deviatoric stress versus axial strain and Figure 5.21 (b) shows the plots of volumetric strain versus axial strain for drained triaxial test.

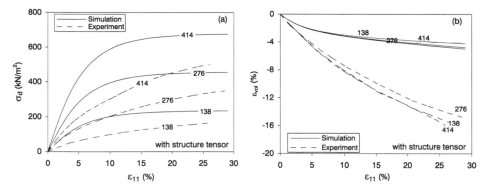

Figure 5.21: Simulation of CID triaxial tests with hypoplasticity with structure tensor; the parameters in Table 5.4 are used; (a) plots of deviator stress versus strain; (b) plots of volumetric strain versus strain

For the CID triaxial tests, the stress-strain curves obtained from the simulation with hypoplasticity plot above the stress-strain curves from the experiments for all three consolidation pressures. In contrast, the volumetric strains from the simulations are much lower than the experimental results.

To apply hypoplasticity with structure tensor to boundary value problems, the initial state of structure stresses must be assigned. The structure stress varies linearly with the consolidation stress; therefore, in the calibration, the relationship between structure stress and initial stress should be determined. Based on the experimental results of Balasubramaniam et al. (1), the relationship between structure stresses and consolidation pressures is shown in Figure 5.22. From the plot, it can be seen that the initial structure stress depends on the stress level in the soil. The initial structure stress increases as the consolidation stress increases.

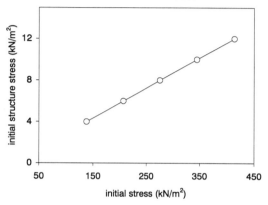

Figure 5.22: Relationship between initial structure stress and isotropic initial stress

5.2.3 Hypoplasticity by von Wolffersdorff

The hypoplasticity by von Wolffersdorff is used to model normally consolidated clay. The calibration of the model for fine-grained soils has been described in Herle (11) and Herle et al. (15), which can be summarized as follows.

1. The critical friction angle φ_c is determined from triaxial test. The friction angle for Bangkok Clay is $26.5°$.

2. h_s and n are calculated from the slope of the normally consolidated line λ. In case of pressure-dependent λ

$$
n = \frac{\ln\left(\dfrac{e_1 \lambda_2}{e_2 \lambda_1}\right)}{\ln\left(\dfrac{p_2}{p_1}\right)} \qquad \text{and} \qquad h_s = 3p\left(\frac{ne}{\lambda}\right)^{1/n}, \tag{5.2}
$$

where 1 and 2 are points at boundaries of the experimental curve, for the calculation of h_s the values of e and λ are taken from the middle of the experimental stress range. For NC-Bangkok Clay, λ_1 is equal to λ_2.

3. e_{i0} is obtained by extrapolation of the normally consolidation line to $p = 0$ by using Eq. 3.20. e_{c0} is obtained by extrapolation of the critical state line to $p = 0$ by using Eq. 3.20. e_{d0} is calculated from the water content at the plastic limit w_p assuming that $p_p = 15$ kN/m^2 (9).

4. For the exponents α and β, typical values for sands can be used (13), e.g. $\alpha = 0.15$ and $\beta = 1.0$. In this research, the trial-and-error is used to obtain these parameters.

h_s (N/m^2)	n	φ_c	e_{c0}	e_{d0}	e_{i0}	α	β
900000	0.28	26.5	4.53	1.98	4.99	0.0	0.8

Table 5.5: Parameters of hypoplasticity version von Wolffersdorff for Bangkok clay

The parameters calibrated from drained triaxial test results (1) are shown in Table 5.5. These parameters are used to simulate triaxial compression tests with different consolidation pressure. Figure 5.23 (a) shows the plots of deviatoric stress versus strain and Figure 5.23 (b) shows the plots of pore pressure versus strain. Figure 5.24 (a) shows the effective stress paths in p', q plots and Figure 5.24 (b) shows the peak stresses in $(e, \log p')$ plot.

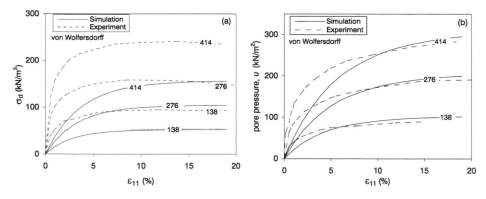

Figure 5.23: Simulations of CIU triaxial tests with hypoplasticity by von Wolffersdorff; the parameters in Table 5.5 are used; (a) plots of deviator stress versus axial strain; (b) plots of pore pressure versus axial strain.

From the simulations of CIU triaxial tests, it can be seen that the hypoplasticity by von Wolffersdorff underpredicts the deviatoric stress for all consolidation pressure. The model underpredicts pore pressures up to $\varepsilon_{11} = 10\%$; however, the model predicts fairly well the pore pressure at limit states for all pressures. The stress paths from the simulation reach the critical state line at lower mean stress than the experiments. This can be seen in the $(e, \log p')$ as well. The simulations of CID triaxial tests are also performed. Figure 5.25 (a) shows the plots of deviatoric stress versus strain and Figure 5.25 (b) shows the plots of pore pressure versus strain. From the CID triaxial tests results, hypoplasticity by von Wolffersdorff predicts the deviator

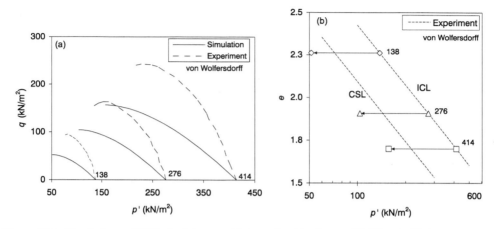

Figure 5.24: Simulations of CIU triaxial tests with hypoplasticity by von Wolffersdorff; the parameters in Table 5.5 are used; (a) stress paths in (q, p') plot; (b) peak stresses in $(e, \log p')$ plot

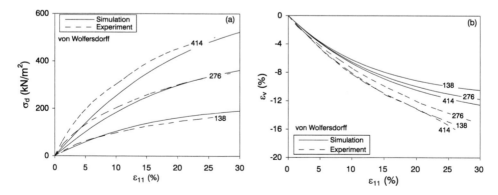

Figure 5.25: Simulations of CID triaxial tests with hypoplasticity by von Wolffersdorff; (a) plots of deviatoric stress versus axial strain; (b) plots of volumetric strain versus axial strain.

stress fairly well for all consolidation stresses. In case of the volumetric strain, the model slightly overpredicts the volumetric strain for all consolidation stresses.

In short, hypoplasticity by von Wolffersdorff is suitable for the drained simulation of NC Bangkok Clay. The model predicts the deviatoric stress and volumetric strain with promising results. The model can not be used to predict undrained behavior of NC Bangkok Clay. The model significantly underpredicts the deviatoric stress and slightly underpredicts the pore pressure.

5.3 Elastoplastic M-C model

The Mohr-Coulomb model is used to model NC Bangkok Clay. The parameters of the model are calibrated from the CIU triaxial test on normally consolidated Bangkok Clay (1). The parameters are used in simulating undrained triaxial test with a consolidation pressure of 276 kN/m^2. Figure 5.26 (a) shows the plots of deviatoric stress versus axial strain and Figure 5.26 (b) shows the plot of pore pressure versus axial strain. Figure 5.27 shows the effective stress paths. From the results

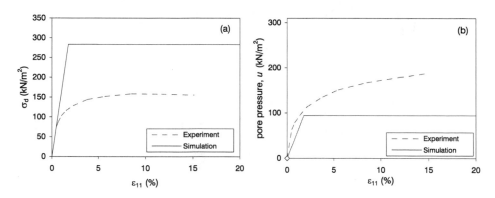

Figure 5.26: Simulation of CIU triaxial test with the M-C model; (a) plots of deviator stress versus axial strain; (b) plots of pore pressure versus axial strain.

of CIU triaxial test, it can be seen that the M-C model significantly overpredicts the deviatoric stress. In contrast, the model underpredicts the pore pressure. The stress path from the simulation reaches the critical state line at higher mean normal stress p' than the stress path of the experiment. In fact, the model shows a constant p' stress path. This is because the mean normal stress in the soil is constant in elastic state. This mean stress does not change until the critical state is reached. Therefore, the deviatoric stress q at the critical state is Mp'_0 which is higher than the deviatoric stress Mp'_f from the experiments.

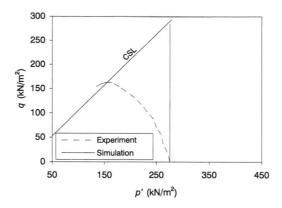

Figure 5.27: Simulation of CIU triaxial test with the M-C model; plot of stress paths

5.4 Comparison of the models

The simulation results of each model at the consolidation stress of 276 kN/m^2 are plotted in the same figure for comparison. Figure 5.28 (a) shows the plots of deviatoric stress versus axial strain and Figure 5.28 (b) shows the plots of pore pressure versus axial strain for the CIU triaxial tests. Figure 5.30 (a) shows the plot of deviatoric stress versus axial strain and Figure 5.30 (b) shows the plot of volumetric strain versus axial strain for the CID triaxial test. From the plots, it can be seen that

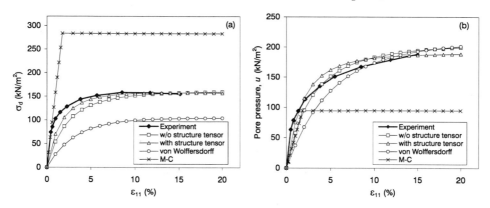

Figure 5.28: Comparison of the simulation results from various models for the CIU triaxial test; consolidation pressure is 276 kN/m^2; (a) plots of deviator stress versus axial strain; (b) plots of volumetric strain versus axial strain

the simulation results obtained by hypoplasticity by Wu with and without structure tensor are close to the experimental results in case of the undrained test. However, the simulation results of the drained case significantly deviate from the experimental results. The hypoplasticity by von Wolffersdorff produces relatively good results for the drained tests. On the other hand, the results obtained with this model show significant discrepancy from the experimental results for undrained condition. For

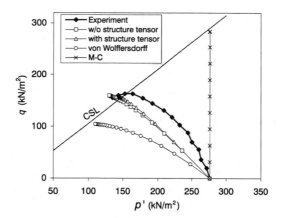

Figure 5.29: Comparison of the simulation results from various models for the CIU triaxial test; the consolidation pressure is 276 kN/m²; plots of effective stress paths

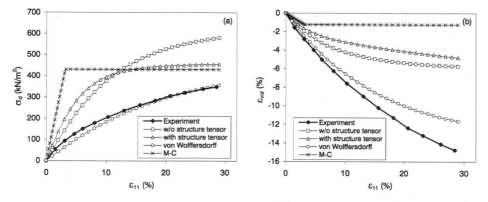

Figure 5.30: Comparison of the simulation results from various models for the CID triaxial test; the consolidation pressure is 276 kN/m²; (a) plots of deviator stress versus axial strain; (b) plots of volumetric strain versus axial strain

the elasto-plastic M-C model, the results obtained with this model also show significant discrepancy from the experimental results for both drained and undrained conditions.

In the next chapter, the EPB shield tunnelling in NC Bangkok clay is simulated. The major concern in the simulation is the behavior of soil at the construction stage. From the discussion above, the hypoplasticity by Wu without structure tensor is selected because the model shows good prediction for the undrained condition. It should be noted that hypoplasticity with structure tensor also gives similar results to the one without structure tensor. However, the hypoplasticity without structure is preferred in the simulation because computing and storing of the structure tensor for every element in FLAC requires more computing time.

Chapter 6

EPB shield tunnelling in Bangkok subsoils

The Chaloem Ratchamongkhon line (MRTA project) is the first subway project which is constructed in Bangkok subsoils to solve the traffic congestion problem. The line is about 20 kilometers long and comprises 18 stations. The route starts from the Hua Lamphong main railway station and ends at the Bang Su railway station. The tunnel axis is located typically at depths between 16 m and 23 m. Because the project route must pass underneath the urban area of Bangkok, the tunnels are required to minimize the disturbance to the ground surface. To predict the amount of settlements, engineers use numerical methods to simulate the tunnelling processes. In this study, the numerical program FLAC is used to simulate the shield tunnelling of the project. The implemented hypoplasticity (as described in 3.3) and the model parameters (as described in 5.2.1) are used in this chapter.

6.1 Geotechnical information

The geological conditions of Bangkok were reported by Muktabhant (36) and Moh et al. (34). Bangkok is located on the Chao Phraya plain which extends from the Gulf of Thailand in the south for a distance about 300 km to the Northwest Highlands. The average width of the plain, from Tanowsri mountain range on the west to the Korat plateau on the east, is about 100 km. The bedrock of the plain is deeper than 300 m. The typical soil profile of Bangkok consists of a 15 m thick layer of soft normally consolidated clay overlying a 5 m thick layer of stiff clay. Below the stiff clay is a thick sand layer down to a depth of 70 m. The layer of normally consolidated clay and the layer of stiff clay are called "Bangkok Clay". The Bangkok Clay originates as a fairly homogeneous and isotropic deposit with no apparent stratification except the stiff zone where small seams of sand are often found. The water content of Bangkok Clay is generally high and close to the liquid limit; however, the water content of the stiff clay drops to the approximate plastic limit.

6.2   EPB shield

The construction of tunnels for the MRTA project by using EPB shield was reported by Chanchaya (3), Muangsaen (35), Maconochie et al. (29) and Teparaksa et al. (45; 46). The tunnelling from Ratchada station to Latprao station is selected to be a case history in this study. Figure 6.1 shows a schematic drawing of the EPB shield used in Ratchada-Latprao section. Table 6.1 shows the dimensions of the EPB shield and tunnel lining. Figure 6.2 shows the typical cross section of tunnels

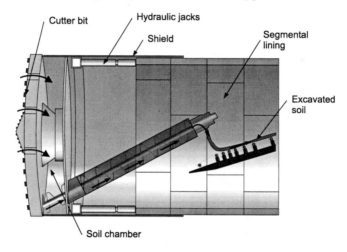

Figure 6.1: Schematic drawing of the EPB shield used in the project

Shield outside diameter	6.43 m
Over-excavation gap	65 mm
Shield overall length	8.35 m
Lining outside diameter	6.30 m
Lining thickness	300 mm
Lining width	1.2 m

Table 6.1: Dimension of EPB shield for Ratchada-Latprao section (29)

at the Ratchada-Latprao section. The northbound and southbound tunnel' axes are located at a depth of 18.1 m below ground surface. The horizontal distance between the axes of the tunnels is 15.1 m. The contractor commenced the EPB shield of the southbound from a launch shaft which is a partitioned section within the Ratchada station box. The northbound EPB shield was started after the southbound tunnel was 100 m away from the launching shaft. The shield is advanced by using cutter bits to cut the soil in front of the tunnel. The excavated soil is converted to mud and collected in the soil chamber. Soil and water pressure at the shield face is resisted by controlling the mud pressure in the soil chamber. The pressure is generated by

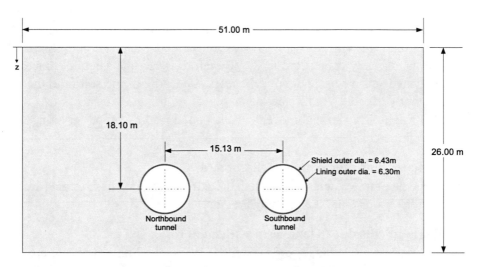

Figure 6.2: Cross section at the monitoring section

propulsion of the hydraulic jacks inside the shield and controlling of soil discharge. As the shield body is advancing, segmental concrete linings are installed to support soil and water pressure. After the shield body has passed, a void between lining and surrounding soil remained. This void was filled by grouting through holes located at the crown segments. In this project, the grouting is done after 2 lining rings (2.4 m) are installed. Typical grouting pressure is 2-3 bar for this section.

Figure 6.3 shows the evolution of settlement along TBM works adopted from Kunito et al. (23) and Lee et al. (28). The figure shows that the ground movements are classified into five categories:

Ground movements ahead and above the face (a) .

This type of ground movements occurs before passing of the EPB shield. It relates to the unbalance between the soil and water pressure in front of the shield and the excavated soil pressure in the soil chamber.

Ground movements along the EPB shield (b) .

This type of ground movements occurs over the shield because of the over-cutting of the soil. More often, the tools at the periphery of the cutting wheel at the face excavate the soil slightly larger than the shield diameter to reduce friction and facilitate the steering of EPB shield. This over-cutting creates a void between soil and the EPB shield. In practice, bentonite slurry is injected into this void to reduce the ground movements.

Ground movements induced at the tail void (c) .

This type of ground movements occurs upon the erection of the tunnel lining. In practice, the outside diameter of the shield must be larger than the outside diameter of the tunnel lining to facilitate the lining installation. This different in diameter will form a void between the excavated soil and lining. This void is normally called "tail void". Therefore, when the shield advances, the excavated soil is unsupported and causes squeezing of the soil in to the void. Generally, the tail void is the major source of the ground movement due to EPB shield. This ground movement can be reduced by proper grouting of the tail void.

Ground movements due to lining deflection (d) .

This type of movements comes from the self-weight and the external pressures form the ground and the grouting pressure.

Ground movements due to long term settlement (e) .

This type of movements occurs after the EPB shield passed and lining is installed due to the direct consequence of the tunnelling and long-term movements due to consolidation of ground around the tunnel.

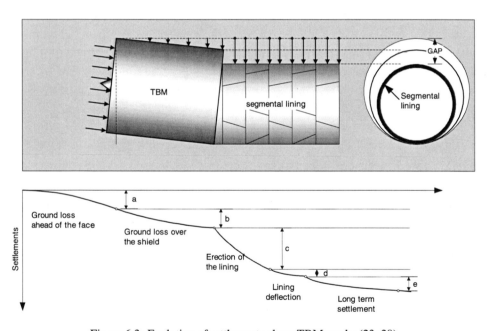

Figure 6.3: Evolution of settlements along TBM works (23; 28)

6.3 Monitoring data

The surface settlement observations are reported by Chanchaya (3). The typical instruments were surface settlements points along and perpendicular to the tunnel axis. Figure 6.4 shows the layout of settlement points arrays located between Ratchada and Latprao station. Array 001 is located at about 103 m from Ratchada station, which corresponding to lining ring number 85. Array 002 is located at distance about 42 m from the station, which corresponding to lining ring number 35.

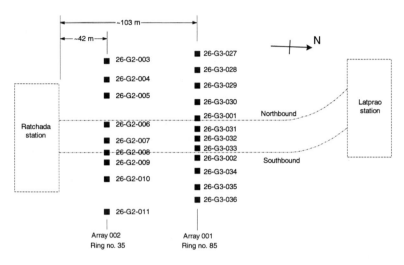

Figure 6.4: Layout of settlement points arrays located between Ratchada and Latprao station

Figure 6.5 shows the settlement profiles monitored from array 001 and array 002. The maximum displacements after the northbound TBM has passed are -40.1 mm and -32.9 mm for array 001 and 002, respectively. The maximum displacements after the southbound TBM has passed are -52.9 mm and -57.8 mm for array 001 and 002, respectively.

Figure 6.6 shows the observed surface displacement along the longitudinal axis of the southbound tunnel when the shield face reaches array 001 and 002. It can be seen that the surface settlements due to the tail void are about 70% and 68% of the total surface settlement for the TBM face at array 001 and 002, respectively.

Muangsaen (35) reported the behavior of EPB shield for the MRTA subway project in Bangkok. The clearances between the outer surface of TBM and outer surface of lining at crown, invert and springline were measured and documented. Figure 6.7 shows the observed clearance between the outer surface of lining and outer surface of the shield. It can be seen that the clearances at the springline and the invert are about 60 mm and the clearance at the crown is about 80 mm. These observed data are used for comparison with the simulation results.

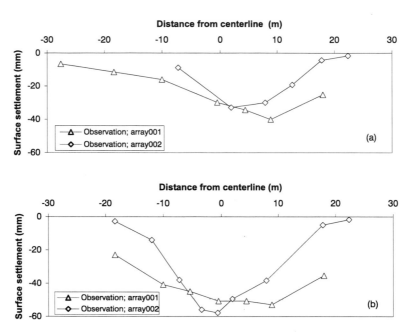

Figure 6.5: Surface settlement profile at selected section (3); (a) after the southbound tunnel has been constructed (b) after the northbound tunnel has been constructed

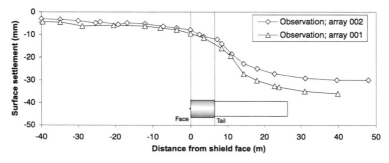

Figure 6.6: Observed surface displacement along the southbound tunnel's axis corresponding to the position of TBM (3) and (35)

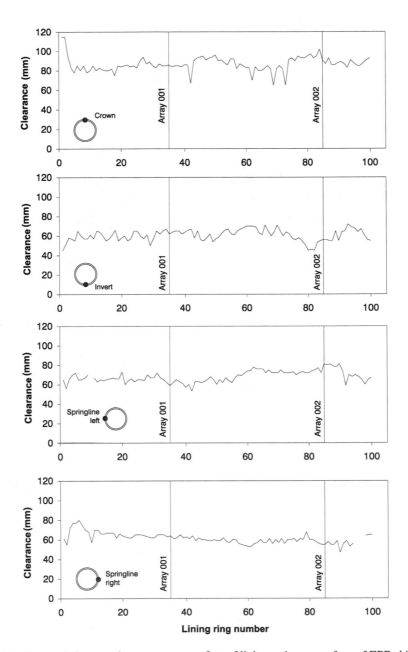

Figure 6.7: Observed clearance between outer surface of lining and outer surface of EPB shield (35)

6.4 Empirical prediction of surface settlements

The ground surface settlement due to tunnelling can be estimated empirically with a method proposed by Peck (39). Peck used the normal probability curve for the estimation of the settlement trough above the tunnel as shown in Figure 6.8. The settlement at any point on the surface can be calculated from

$$\delta = \delta_{max} \exp\left(-\frac{x^2}{2i^2}\right),\tag{6.1}$$

where δ_{max} is the maximum surface settlement which is empirically determined, i is the horizontal distance from tunnel axis to the point of inflection of the curve. The δ_{max} is estimated from the amount of volume loss V_u/A proposed by Mair et al. (30). For an supported face tunnelling in soft clay, $V_u/A = 2\%$ is used. The δ_{max} is calculated from the area of the probability curve

$$\delta_{max} = \frac{V_u}{\sqrt{2\pi}i}\tag{6.2}$$

The i value is calculated by using a dimensionless plot of i/R against $z/2R$, as shown in Figure 6.8. If the two tunnels of a pair are close enough to produce a single settlement trough, it may be interpreted as a single tunnel with dept z and radius $R' = R + d/2$ (39), where d is the horizontal distance between the axes of the tunnels and R is the radius of the single tunnel. The i value is calculated by using the plot between $z/2R'$ versus i/R' (as shown in Figure 6.8). The results of the prediction with this method are shown in Section 6.6.

6.5 Numerical simulations of EPB shield tunnelling

FLAC is used to simulate EPB shield tunnelling. The soil is discretized into small grid zones (as described in 2.3). Figure 6.9 shows the discretized mesh. The simulation is done according to the construction sequences. The southbound tunnel is excavated first followed by the northbound tunnel. In this study, the term "stage I" is used for the simulation of the southbound tunnel and the term "stage II" is used for the simulation of the northbound tunnel. The initial state of stress and pore water pressure are applied to all grid zones of the mesh. The hydrostatic pore pressure is assumed in the simulation.

The gap dimension due to the difference between excavation boundary and lining is taken from the observation during construction (Fig.6.7). The movements due to ground loss ahead and over the shield which are tree-dimensional movements are

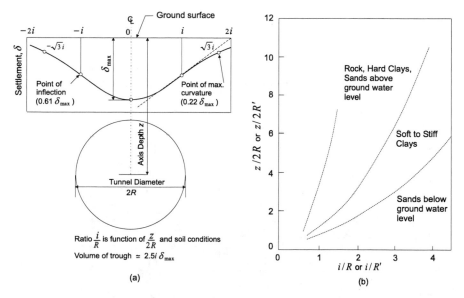

Figure 6.8: (a) Properties of the normal probability curve as used to represent settlement trough above tunnel; (b) relation between dimensionless width of settlement trough i/R and dimensionless depth of tunnel $z/2R$ for various tunnels in different materials. (39)

approximately added in the two-dimensional model by assuming a slightly larger excavation diameter (42). In this study, the settlement due to ground loss ahead of the face and over the shield is about 30% of the settlement due to the tail void.

The excavation process is modelled based on the method by Swoboda (43). All grid zones enclosed by tunnel's boundary are removed . The nodes on the tunnel boundary are fixed for the calculations of out-of-balance nodal forces due to the removing of the zones. After the out-of-balance nodal forces are obtained, these nodal forces are applied back in the opposite direction to maintain equilibrium. These nodal forces are gradually reduced. Segmental linings are modelled with beam elements. The connections between the linings are assumed to be continuous. The void between the excavated soil and the lining is modelled by generating mesh into actual dimensions. The boundary of the excavated soil and beam element are connected by interfaces (17). Large strain mode is used to simulate actual intrusion of the soil into the void. Soil load and water pressure are transferred to the lining when the mesh contacts with the beam elements. The soil is modelled with hypoplasticity without structure tensor which calibrated with CIU triaxial test described in 5.2.1. The parameters are shown in Table 5.3.

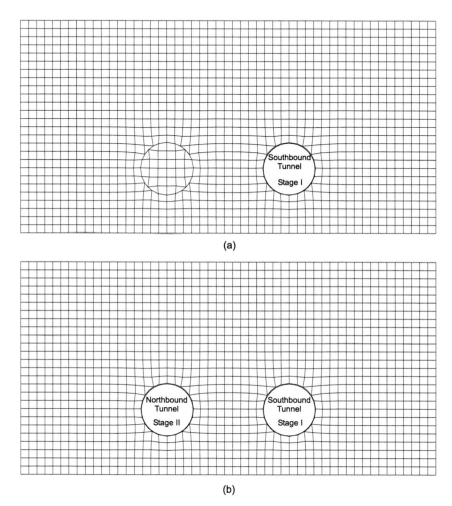

(a)

(b)

Figure 6.9: Mesh discretization for EPB shield tunnelling; (a) Excavate northbound tunnel (stage I) ; (b) excavate southbound tunnel (stage II)

6.6 Results

6.6.1 Influence of K_0

This section describes the influence of K_0 on the prediction of ground surface settlement. Niemunis (37) reported that the earth pressure coefficient at rest K_0 cannot be independently chosen for hypoplastic model. For normally consolidated soil, the value of K_0 is higher than the one obtained with Jaky's equation ($K_0 = 1 - \sin \varphi$). The upper limit of K_0 can be approximated with visco-hypoplasticity from a constant-strain-rate (CRSN) oedometric test with OCR=1 and $\mathbf{D} \sim \mathbf{D}^{vis}$; where $\mathbf{D}$ is stretching and $\mathbf{D}^{vis}$ is viscous rate. The equation for the approximation of K_0^{up} reads

$$K_0^{up} = \frac{-2 - a^2 + \sqrt{36 + 36a^2 + a^4}}{16}.$$ (6.3)

Parameter a is calculated from Equation 3.22. Figure 6.10 shows the plots of K_0 versus φ calculated from the Jaky's equation and from Equation 6.3. It can be seen that

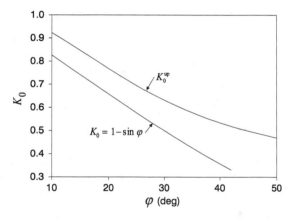

Figure 6.10: The values of K_0^{up} obtained from the condition of uniaxial creep for normally consolidated materials (37)

K_0 calculated with visco-hypoplasticity for normally consolidated material is higher than the one calculated with Jaky's equation. The parametric study is performed for determining of an appropriate K_0. K_0 is varied from 0.6 to 0.8 in the simulation with FLAC. Figure 6.11 shows the plots of ground surface settlement obtained with FLAC for different K_0. For $K_0 = 0.6$ the settlement trough shows a heave of 20 mm near the lateral boundary. The heave reduces as K_0 increases. The width of settlement trough increase as K_0 increases. From this parametric study, $K_0 = 0.8$ is used in this study as it gives a reasonable prediction of the ground surface settlements.

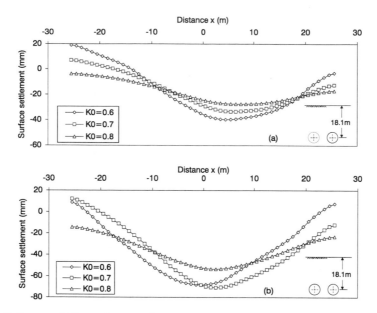

Figure 6.11: Influence of K_0 on the ground surface settlement; (a) after the passing of the southbound EPB shield, "stage I"; (b) after the passing of northbound EPB shield, "stage II"

6.6.2 Comparison of prediction and observation

Figure 6.12 shows the plots of surface settlements along the cross section obtained with FLAC. Figure 6.13 shows the plots of surface settlements obtained with the empirical method by Peck. The observed settlements are plotted in each figure for comparison. From the results, it can be seen that FLAC underpredicts the ground surface settlements for stage I. The width of the settlement trough obtained with FLAC is wider than the observations. For stage II, FLAC results agree well with the observation both the maximum settlement and the shape of the settlement trough. In case of the empirical method, for stage I the width of the settlement trough is narrower than the observations. For stage II, the empirical method significantly underpredicts the maximum surface settlement compared with the observations and the results from FLAC.

It can be seen that, for the empirical method, the maximum surface settlement δ_{max} must be determined empirically, as described in Section 6.4. Therefore, the results strongly depend on the estimation of volume loss and the i value which are based solely on experience. In contrast, FLAC does not need the predefined maximum surface settlement δ_{max}. The actual dimensions of tunnels are used to generate the mesh directly. As a result, the geometries of the tunnel, the lining and the tail void are modelled without any predefined maximum surface settlement. However, for the two-dimensional simulation with FLAC, the tree-dimensional soil movement ahead and over the shield must be assumed. In fact, if the simulations are done with three-dimensional version of FLAC, the assumption of the soil movement at the shield

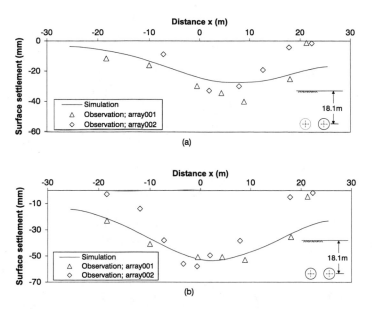

Figure 6.12: Comparison of surface settlement profiles obtained with FLAC and observation; (a) after the passing of the southbound EPB shield, "stage I"; (b) after the passing of northbound EPB shield, "stage II"

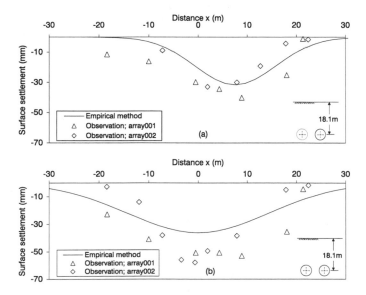

Figure 6.13: Comparison of surface settlement profiles obtained with the empirical method by Peck (39) and observation; the volume loss Vu/A is estimated from the suggestion by Mair et al. (30); (a) after the passing of the southbound EPB shield, "stage I"; (b) after the passing of northbound EPB shield, "stage II"

face can be eliminated, which leads to more accurate simulation.

Chapter 7

Conclusions and recommendations

The influences of the material nonlinearity, the mass density and the artificial damping on the numerical solutions of the finite difference FLAC were investigated. The author found that material nonlinearity provides some damping to the system. Nodal accelerations reduce significantly if nonlinear constitutive equations such as hypoplasticity or hyperbolic elasticity are used. The inertia of nodal points is increased by increasing the nodal mass. Nodal accelerations reduce as the nodal masses increase. The artificial damping is added for speeding the convergence for quasi-static problems. The damping factor of 0.8 was used in this study.

The hypoplastic constitutive equations by Wu with and without structure tensor and by von Wolffersdorff were implemented in FLAC. The implementations were verified with element tests using only one element. For the hypoplastic constitutive equations by Wu and by von Wolffersdorff the results obtained with FLAC were similar to the "exact" solution. For hypoplasticity with structure tensor, the solutions slightly differed from the "exact" solutions. The implemented code of hypoplasticity by Wu was verified by the simulation of some laboratory tests and some geotechnical problems. Shear bands formations were found in the numerical simulations of biaxial test and simple shear test which were close to shear bands obtained from experiments.

For applications to geotechnical problems, the bearing capacity factors N_γ of a spread footing calculated with hypoplasticity by Wu were close to the analytical solution. However, the elasto-plasticity with Mohr-Coulomb yield criterion underpredicted N_γ. It should be noted that for non-dilatant hypoplasticity by Wu ($\psi = 0°$), an instability occurred when applied to the spread footing problem . The instability reduces as ψ approaches to φ. However, this problem was not found for unloading problems such as trapdoor or tunnelling problems. For the trapdoor problem, the limit stress ratio obtained by FLAC with hypoplasticity by Wu is close to the analytical solution. The shape of the average stress profiles over the trapdoor are also similar to the experimental results. For the simulation of an unlined circular tunnel, the mesh arrangement does not show significant effect on the numerical results.

The hypoplastic constitutive equations by Wu with and without structure tensor, by von Wolffersdorff and M-C model were used to model NC Bangkok Clay. The hypoplasticity by Wu with and without structure tensor predicted undrained behavior quite well but it cannot be used to model drained behavior. Hypoplasticity by von Wolffersdorff predicted well the drained behavior; however, the model did not give good prediction for the undrained case. Therefore, the hypoplasticity by Wu without structure tensor was selected to model NC Bangkok Clay under undrained condition.

The hypoplasticity by Wu was used to model NC Bangkok Clay for shield tunnelling in Bangkok. The tail void was modelled by using actual dimensions from the construction records. The ground surface settlements from the simulations were close to the observation data. In contrast, the empirical method underpredicted the ground surface settlements.

Further studies are needed for the following topics:

- double precision version of FLAC2D should be used to investigate the influence of numerical precision on the numerical results,

- current versions of hypoplasticity which is able to model creep and relaxation (40; 10; 37) should be used to model NC Bangkok Clay,

- effects of three dimensional deformations ahead of shield face should be examined by using FLAC3D. The implemented code for FLAC2D can be used for FLAC3D with trivial adjustment,

- effect of grouting on the ground surface settlement should be investigated.

Bibliography

[1] A. S. Balasubramaniam and A. R. Chaudhry. Deformation and strength characteristics of soft Bangkok Clay. *Journal of the Geotechnical Engineering Division, ASCE*, 104:1153–1167, 1978.

[2] M. Budhu. *Soil Mechanics and Foundations*. John Wiley and Sons Inc., 2000.

[3] C. Chanchaya. Analysis of movements in Bangkok subsoils due to subway tunnelling by the soil modelling method. Master's thesis, Chulalongkorn University, 2000.

[4] P. A. Cundall. *Analytical and Computational Methods in Engineering Rock Mechanics*. Allen & Unwin, London, 1987.

[5] R. de Borst and P. A. Vermeer. Possibilities and limitations of finite elements for limit analysis. *Geotechnique*, 34(2):199–210, 1984.

[6] W. Fellin and A. Ostermann. Consistent tangent operators for constitutive rate equations. *Int. J. for Numer. Anal. Meth. Geomech.*, 26:1213–1233, 2002.

[7] D. Frydman and J. H. Burd. Numerical studies of bearing-capacity factor n_γ. *Journal of the Geotechnical Engineering Division, ASCE*, 123(1):20–29, 1995.

[8] D. V. Griffiths. Computation of bearing capacity factors using finite elements. *Geotechnique*, 32(3):195–202, 1982.

[9] G. Gudehus. A comprehensive constitutive equation for granular material. *Soils and Foundation*, 36(1):1–12, 1996.

[10] G. Gudehus. A visco-hypoplastic constitutive relation for soft soils. *Soils and Foundation*, 44(4):11–25, 2004.

[11] I. Herle. Granulometric limits of hypoplastic models. *TASK Quarterly*, 4(3):389–407, 2000.

[12] I. Herle. Constitutive models for numerical simulations. In D. Kolymbas, editor, *Summerschool, Innsbruck*. Logos, 2001.

[13] I. Herle and G. Gudehus. Determination of parameters of a hypoplastic constitutive model from properties of grain assemblies. *Mechanics of Cohesive-Frictional Materials*, 4(5):461–486, 1999.

[14] I. Herle and Feda J. Interaction of spread footing with sandy subsoil. Part 2: Plane strain FE modelling. *Engineering Mechanics*, 9(4):259–272, 2002.

[15] I. Herle and D. Kolymbas. Hypoplastic for soils with low friction angles. *Computers and Geotechnics*, 31(5):365–373, 2004.

[16] H. Hügel. Prognose von Bodenverformungen. Veröffentlichung des Institutes für Bodenmechanik und Felsmechanik der Universität Fridericiana in Karlsruhe, Heft 136, 1995.

[17] Itasca. *FLAC (Fast Lagrangian Analysis of Continua)*. Itasca Consulting Group, Inc., Minnesota, U.S.A., 2000.

[18] R. G. James and A. S. Balasubramaniam. The peak stress envelopes and their relations to the critical state line for a saturated clay. *Proceeding of 4th Asian Regional Conference on Soil Mechanics and Foundation, Bangkok, Thailand*, pages 115–120, 1971.

[19] D. Kolymbas. *Introduction to Hypoplasticity*. Advances in Geotechnical Engineering and Tunnelling. Balkema, 2000.

[20] D. Kolymbas. Tunnelling mechanics. In D. Kolymbas, editor, *Eurosummerschool, Innsbruck*. Logos, 2001.

[21] D. Kolymbas, M. Mähr, and P. Tanseng. Some principles for the design of lining. In D. Kolymbas, editor, *Summerschool, Innsbruck*. Logos, 2003.

[22] N. C. Koutsabeloulis and D. V. Griffiths. Numerical modelling of the trapdoor problem. *Geotechnique*, 39(1):77–89, 1989.

[23] S. Kunito and N. B. Sugden. Control of ground movements due to tunelling with an earth pressure balance TBM. In *Proceedings of Rapid Excavation and Tunneling Conference, Littleton, Colorado*, pages 129–139. Society for Mining, Metallurgy and Exploration, Inc., 2001.

[24] C. C. Ladd and R. Foott. New design procedure for stability of soft clays. *Journal of the Geotechnical Engineering Division, ASCE*, 100(GT7):763–786, 1974.

[25] C. C. Ladd, R. Foott, K. Ishihara, F. Schlosser, and H. G. Poulos. Stress-deformation and strength characteristic. *Proceeding of 9th International Conference on Soil Mechanics and Foundation Engineering, Tokyo.*, 2:421–494, 1977.

[26] P. V. Lade and H. M. Musante. Failure conditions in sand and remoulded clay. *Proceeding of 9th International Conference on Soil Mechanics and Foundation Engineering, Tokyo.*, 1:181–186, 1977.

[27] P. V. Lade and H. M. Musante. Three-dimensional behavior of remolded clay. *Journal of the Geotechnical Engineering Division, ASCE*, 104:193–209, 1978.

[28] K. M. Lee, R. K. Rowe, and K. Y. Lo. Subsidence owing to tunnelling. I. estimating the gap parameter. *Canadian Geotechnical Journal*, 29:929–940, 1992.

[29] D. J. Maconochie, S. Suwansawat, and C. C. Chang. Tunneling for the Chaloem Ratchamongkhon line in Bangkok. In *Proceedings of Rapid Excavation and Tunneling Conference, Littleton, Colorado*, pages 113–128. Society for Mining, Metallurgy and Exploration, Inc., 2001.

[30] R. J. Mair and R. N. Taylor. Bored tunnelling in the urban environment. In *Proceedings of the 14th International Conference on Soil Mechanics and Foundation Engineering*, Hamburg, 1997.

[31] N. Manoharan and S. P. Dasgupta. Bearing capacity of surface footings by finite elements. *Computers and Structures*, 54(4):563–586, 1995.

[32] Th. Marcher, P.A. Vermeer, and von Wolffersdorff P.-A. Hypoplastic and elastoplastic modelling – a comparison with test data. In D. Kolymbas, editor, *Constitutive Modelling of Granular Materials*. Springer, 2000.

[33] R. D. Mindlin. Stress distribution around a tunnel. In *ASCE Proceedings*, pages 619–649, April 1939.

[34] Z. C. Moh, J. D. Nelson, and E. W. Brand. Strength and deformation behavior of Bangkok Clay. In *Proceedings of the 7th International Conference on Soil Mechanics and Foundation Engineering*, volume 1, 1969.

[35] K. Muangsaen. Behavior of EPB shield for subway construction in Bangkok subsoil. Master's thesis, Chulalongkorn University, 2001.

[36] C. Muktabhant. Engineering properties of Bangkok subsoil. In *Proceedings Southeast Asian Regional Conference on Soil Engineering*, pages 1–7, 1967.

[37] A. Niemunis. Extended hypoplastic models for soils. Schriftenreihe des Institutes für Grundbau und Bodenmechanik der Ruhr-Universität Bochum, Heft 34, 2003.

[38] E. Papamichos, I. Vardoulakis, and L. K. Heil. Overburden modeling above a compacting reservoir using a trap door apparatus. *Phys. Chem. Earth (A)*, 26(1–2):69–74, 2001.

[39] R. B. Peck. Deep excavations and tunneling in soft ground. State-of-the-Art report. In *Proceedings of the 7th International Conference on Soil Mechanics and Foundation Engineering*, volume State-of-the-Art Volume, pages 225–290, Mexico City, 1969.

[40] A. Punlor. Numerical modelling of the visco-plastic behavior of soft soils. Veröffentlichung des Institutes für Bodenmechanik und Felsmechanik der Universität Fridericiana in Karlsruhe, Heft 163, 2004.

[41] A. Ph. Revuzhenko. *Mechanika synushej sredj (Mechanics of the loose medium)*. ОФСЕТ publishing, Novosibirsk, Russia, 2003.

[42] R. K. Rowe and G. J. Kack. A theoretical examination of the settlements induced by tunnelling: four case histories. *Canadian Geotechnical Journal*, 20:299–314, 1983.

[43] G. Swoboda. Numerical modelling of tunnels. In C.S. Desai and G. Gioda, editors, *Numerical Methods and Constitutive Modelling in Geomechanics*. Springer, 1990.

[44] J. Tejchman. Modelling of shear localisation and autogeneous dynamic effects in granular bodies. Veröffentlichung des Institutes für Bodenmechanik und Felsmechanik der Universität Fridericiana in Karlsruhe, Heft 140, 1997.

[45] W. Teparaksa and C. Chanchaya. Analysis of movements in Bangkok subsoils due to subway tunnelling. In *Proceedings of the 7th National Convention on Civil Engineering*, Thailand, 2001.

[46] W. Teparaksa and W. Pitaksaithong. Behavior and prediction of soil displacement during (tbm) tunnelling in Bangkok subsoils. In *Proceedings of the 7th National Convention on Civil Engineering*, Thailand, 2001.

[47] K. Terzaghi. Stress distribution in dry and saturated sand above a yielding trapdoor. In *Proceedings of International Confernce on Soil Mechanics*, volume 1, pages 307–311, Cambridge Mass., 1936.

[48] I. Vardoulakis. Scherfugenbildung in sandk"orpern als verzweigungsproblem. Veröffentlichung des Institutes für Bodenmechanik und Felsmechanik der Universität Fridericiana in Karlsruhe, Heft 70, 1977.

[49] I Vardoulakis, B. Graf, and G. Gudehus. Trap-door problem with dry sand: A statical approach based upon model test kinematics. *Int. J. for Numer. Anal. Meth. Geomech.*, 5:57–78, 1981.

[50] P. A. Vermeer and R. de Borst. Non-associated plasticity for soils, concrete and rock. *Heron*, 29(3), 1984.

[51] P.-A. von Wolffersdorff. A hypoplastic relation for granular materials with a predefined limit state surface. *Mechanics of Cohesive-Frictional Materials*, 1:251–271, 1996.

[52] P. K. Woodward and D. V. Griffiths. Observations on the computation of the bearing capacity factor n_γ by finite elements. *Geotechnique*, 34(2):199–210, 1984.

[53] W. Wu and D. Kolymbas. Numerical testing of the stability criterion for hypoplastic constitutive equations. *Mechanics of Materials*, 9:245–253, 1990.

[54] W. Wu and D. Kolymbas. Hypoplasticity then and now. In D. Kolymbas, editor, *Constitutive Modelling of Granular Materials*. Springer, 2000.

[55] W. Wu and O. P. Roony. The role of numerical analysis in tunnel design. In D. Kolymbas, editor, *Eurosummerschool, Innsbruck*. Springer, 2001.

[56] O. C. Zienkiewicz, C. Humpheson, and R. W. Lewis. Associated and non-associated visco-plasticity and plasticity in soil mechanics. *Geotechnique*, 25(4):671–689, 1975.

Appendix A

Calibration of hypoplasticity

A.1 Calibration with drained triaxial test

Four parameters are required for the hypoplastic constitutive equation, viz. C_1, C_2, C_3 and C_4; as a result, four known values from the test results, viz. E_i, β_A, β_B and $(\sigma_1 - \sigma_2)_{max}$ (see Fig. 5.11) are used in the calibration.

The strain increment ($\Delta\varepsilon$) at point A and B can be determined from β_A and β_B as follows:

$$\tan^{-1}\left(\frac{\Delta\varepsilon_{11} + 2\Delta\varepsilon_{22}}{\Delta\varepsilon_{11}}\right)_A = \beta_A \tag{A.1}$$

$$\tan^{-1}\left(\frac{\Delta\varepsilon_{11} + 2\Delta\varepsilon_{22}}{\Delta\varepsilon_{11}}\right)_B = \beta_B$$

Because the hypoplastic constitutive equation is rate-independent, we can set $\Delta\varepsilon_{11} = -1$. Then $\Delta\varepsilon_{22}$ and $\Delta\varepsilon_{33}$ are calculated from

$$\Delta\varepsilon_{22} = \Delta\varepsilon_{33} = \frac{1}{2}(1 - \tan\beta) \tag{A.2}$$

The strain increment tensors are

$$\Delta\boldsymbol{\varepsilon}_A = \begin{pmatrix} -1 & 0 & 0 \\ 0 & \frac{1}{2}(1 - \tan\beta_A) & 0 \\ 0 & 0 & \frac{1}{2}(1 - \tan\beta_A) \end{pmatrix}$$

$$\Delta\boldsymbol{\varepsilon}_B = \begin{pmatrix} -1 & 0 & 0 \\ 0 & \frac{1}{2}(1 - \tan\beta_B) & 0 \\ 0 & 0 & \frac{1}{2}(1 - \tan\beta_B) \end{pmatrix}$$

The stress tensors at point A and B are

$$\boldsymbol{\sigma}_A = \begin{pmatrix} \sigma_{11A} & 0 & 0 \\ 0 & \sigma_{22A} & 0 \\ 0 & 0 & \sigma_{33A} \end{pmatrix} = \begin{pmatrix} \sigma_c & 0 & 0 \\ 0 & \sigma_c & 0 \\ 0 & 0 & \sigma_c \end{pmatrix}$$

$$\boldsymbol{\sigma}_B = \begin{pmatrix} \sigma_{11B} & 0 & 0 \\ 0 & \sigma_{22B} & 0 \\ 0 & 0 & \sigma_{33B} \end{pmatrix} = \begin{pmatrix} \sigma_{11B} & 0 & 0 \\ 0 & \sigma_c & 0 \\ 0 & 0 & \sigma_c \end{pmatrix}$$

where σ_c is the cell pressure which is constant for the conventional triaxial test. The value of σ_{11B} is calculated from the friction angle and the cell pressure σ_c as follows

$$\sigma_{11B} = \sigma_c \left(\frac{1 + \sin\varphi}{1 - \sin\varphi} \right) \tag{A.3}$$

The incremental stress tensors $\Delta\boldsymbol{\sigma}_A$ and $\Delta\boldsymbol{\sigma}_B$ are

$$\Delta\boldsymbol{\sigma}_A = \begin{pmatrix} -E & 0 & 0 \\ 0 & 0 & 0 \\ 0 & 0 & 0 \end{pmatrix}, \quad \Delta\boldsymbol{\sigma}_B = \begin{pmatrix} 0 & 0 & 0 \\ 0 & 0 & 0 \\ 0 & 0 & 0 \end{pmatrix}$$

Finally, the parameters C_1, C_2, C_3 and C_4 are calculated by solving the system of equations

$$\begin{pmatrix} \text{tr}(\boldsymbol{\sigma}_A)\Delta\varepsilon_{11A} & \dfrac{\text{tr}(\boldsymbol{\sigma}_A\Delta\boldsymbol{\varepsilon}_A)}{\text{tr}(\boldsymbol{\sigma}_A)}\sigma_{11A} & \dfrac{\sqrt{\text{tr}(\Delta\boldsymbol{\varepsilon}_A^2)}}{\text{tr}(\boldsymbol{\sigma}_A)}\sigma_{11A}^2 & \dfrac{\sqrt{\text{tr}(\Delta\boldsymbol{\varepsilon}_A^2)}}{\text{tr}(\boldsymbol{\sigma}_A)}(\sigma_{11A}^*)^2 \\[3mm] \text{tr}(\boldsymbol{\sigma}_A)\Delta\varepsilon_{33A} & \dfrac{\text{tr}(\boldsymbol{\sigma}_A\Delta\boldsymbol{\varepsilon}_A)}{\text{tr}(\boldsymbol{\sigma}_A)}\sigma_{33A} & \dfrac{\sqrt{\text{tr}(\Delta\boldsymbol{\varepsilon}_A^2)}}{\text{tr}(\boldsymbol{\sigma}_A)}\sigma_{33A}^2 & \dfrac{\sqrt{\text{tr}(\Delta\boldsymbol{\varepsilon}_A^2)}}{\text{tr}(\boldsymbol{\sigma}_A)}(\sigma_{33A}^*)^2 \\[3mm] \text{tr}(\boldsymbol{\sigma}_B)\Delta\varepsilon_{11B} & \dfrac{\text{tr}(\boldsymbol{\sigma}_B\Delta\boldsymbol{\varepsilon}_A)}{\text{tr}(\boldsymbol{\sigma}_B)}\sigma_{11B} & \dfrac{\sqrt{\text{tr}(\Delta\boldsymbol{\varepsilon}_B^2)}}{\text{tr}(\boldsymbol{\sigma}_B)}\sigma_{11B}^2 & \dfrac{\sqrt{\text{tr}(\Delta\boldsymbol{\varepsilon}_B^2)}}{\text{tr}(\boldsymbol{\sigma}_B)}(\sigma_{11B}^*)^2 \\[3mm] \text{tr}(\boldsymbol{\sigma}_B)\Delta\varepsilon_{33B} & \dfrac{\text{tr}(\boldsymbol{\sigma}_B\Delta\boldsymbol{\varepsilon}_B)}{\text{tr}(\boldsymbol{\sigma}_B)}\sigma_{33B} & \dfrac{\sqrt{\text{tr}(\Delta\boldsymbol{\varepsilon}_B^2)}}{\text{tr}(\boldsymbol{\sigma}_B)}\sigma_{33B}^2 & \dfrac{\sqrt{\text{tr}(\Delta\boldsymbol{\varepsilon}_B^2)}}{\text{tr}(\boldsymbol{\sigma}_B)}(\sigma_{33B}^*)^2 \end{pmatrix} \begin{pmatrix} C_1 \\ C_2 \\ C_3 \\ C_4 \end{pmatrix} =$$

$$= \begin{pmatrix} -E_i \\ 0 \\ 0 \\ 0 \end{pmatrix} \tag{A.4}$$

A.2   Calibration with undrained triaxial test

For the hypoplastic constitutive equation, which has four parameters, the four known values must be determined from experimental results. In case of undrained triaxial test, known values are initial tangent modulus, E_i, maximum deviatoric stress, $(\sigma_1 - \sigma_3)_{\text{max}}$, initial slope of pore pressure-strain curve, $\Delta u_i / \Delta\varepsilon_1$, and the maximum pore pressure, u_{max}. For undrained conditions, the volume of the soil sample is constant throughout the test, i.e.

$$\Delta\varepsilon_v = \Delta\varepsilon_{11} + \Delta\varepsilon_{22} + \Delta\varepsilon_{33} = 0 \tag{A.5}$$

For conventional triaxial test, $\Delta\varepsilon_{22} = \Delta\varepsilon_{33}$; hence

$$\Delta\varepsilon_{11} + 2\Delta\varepsilon_{33} = 0 \tag{A.6}$$

As the hypoplastic constitutive equation is rate-independent, the value of $\Delta\varepsilon_1$ is not important. Therefore, the value of $\Delta\varepsilon_1$ is set to be -1. It then follows

$$\Delta\varepsilon_{33} = 0.5 \tag{A.7}$$

For the undrained case, the tensors $\Delta\varepsilon$ at points A and B are equal

$$\Delta\varepsilon_A = \Delta\varepsilon_B = \begin{pmatrix} -1 & 0 & 0 \\ 0 & 0.5 & 0 \\ 0 & 0 & 0.5 \end{pmatrix}$$

The effective stress tensors σ' at points A and B are

$$\sigma'_A = \begin{pmatrix} \sigma'_{11A} & 0 & 0 \\ 0 & \sigma'_{22A} & 0 \\ 0 & 0 & \sigma'_{33A} \end{pmatrix} = \begin{pmatrix} \sigma_c - u_A & 0 & 0 \\ 0 & \sigma_c - u_A & 0 \\ 0 & 0 & \sigma_c - u_A \end{pmatrix}$$

$$\sigma'_B = \begin{pmatrix} \sigma'_{11B} & 0 & 0 \\ 0 & \sigma'_{22B} & 0 \\ 0 & 0 & \sigma'_{33B} \end{pmatrix} = \begin{pmatrix} \sigma_{1B} - u_B & 0 & 0 \\ 0 & \sigma_c - u_B & 0 \\ 0 & 0 & \sigma_c - u_B \end{pmatrix}$$

where σ_c is a cell pressure which is constant for the conventional triaxial test; and u_A and u_B are the pore pressures at point A and at point B respectively. The value of σ'_{1B} is related to the friction angle by

$$\sigma'_{1B} = \sigma'_{3B} \left(\frac{1 + \sin\varphi'}{1 - \sin\varphi'} \right) \tag{A.8}$$

Finally, if we set $\Delta\varepsilon_1 = -1$, we know the stress increment $\Delta\sigma'_{33A} = \Delta u_i$ and $\Delta\sigma'_{11A} = -E_i + \Delta u_i$; the incremental stress tensors $\Delta\sigma'_A$ and $\Delta\sigma'_B$ are

$$\Delta\sigma'_A = \begin{pmatrix} -E_i + \Delta u_i & 0 & 0 \\ 0 & \Delta u_i & 0 \\ 0 & 0 & 0 \end{pmatrix}, \quad \Delta\sigma'_B = \begin{pmatrix} 0 & 0 & 0 \\ 0 & 0 & 0 \\ 0 & 0 & 0 \end{pmatrix}$$

From the known values above, the parameters $C_1, C_2, C_3,$ and C_4 can be calculated by solving the following relationship

$$
\begin{pmatrix}
\mathrm{tr}(\sigma'_A)\Delta\varepsilon_{11A} & \dfrac{\mathrm{tr}(\sigma'_A\Delta\varepsilon_A)}{\mathrm{tr}(\sigma'_A)}\sigma'_{11A} & \dfrac{\sqrt{\mathrm{tr}(\Delta\varepsilon_A^2)}}{\mathrm{tr}(\sigma'_A)}\sigma'^{2}_{11A} & \dfrac{\sqrt{\mathrm{tr}(\Delta\varepsilon_A^2)}}{\mathrm{tr}(\sigma'_A)}(\sigma'^{*}_{11A})^2 \\[2.2em]
\mathrm{tr}(\sigma'_A)\Delta\varepsilon_{33A} & \dfrac{\mathrm{tr}(\sigma'_A\Delta\varepsilon_A)}{\mathrm{tr}(\sigma'_A)}\sigma'_{33A} & \dfrac{\sqrt{\mathrm{tr}(\Delta\varepsilon_A^2)}}{\mathrm{tr}(\sigma'_A)}\sigma'^{2}_{33A} & \dfrac{\sqrt{\mathrm{tr}(\Delta\varepsilon_A^2)}}{\mathrm{tr}(\sigma'_A)}(\sigma'^{*}_{33A})^2 \\[2.2em]
\mathrm{tr}(\sigma'_B)\Delta\varepsilon_{11B} & \dfrac{\mathrm{tr}(\sigma'_B\Delta\varepsilon_A)}{\mathrm{tr}(\sigma'_B)}\sigma'_{11B} & \dfrac{\sqrt{\mathrm{tr}(\Delta\varepsilon_B^2)}}{\mathrm{tr}(\sigma'_B)}\sigma'^{2}_{11B} & \dfrac{\sqrt{\mathrm{tr}(\Delta\varepsilon_B^2)}}{\mathrm{tr}(\sigma'_B)}(\sigma'^{*}_{11B})^2 \\[2.2em]
\mathrm{tr}(\sigma'_B)\Delta\varepsilon_{33B} & \dfrac{\mathrm{tr}(\sigma'_B\Delta\varepsilon_B)}{\mathrm{tr}(\sigma'_B)}\sigma'_{33B} & \dfrac{\sqrt{\mathrm{tr}(\Delta\varepsilon_B^2)}}{\mathrm{tr}(\sigma'_B)}\sigma'^{2}_{33B} & \dfrac{\sqrt{\mathrm{tr}(\Delta\varepsilon_B^2)}}{\mathrm{tr}(\sigma'_B)}(\sigma'^{*}_{33B})^2
\end{pmatrix}
\begin{pmatrix} C_1 \\ C_2 \\ C_3 \\ C_4 \end{pmatrix} =
$$

$$
= \begin{pmatrix} -E_i + \Delta u_i \\ \Delta u_i \\ 0 \\ 0 \end{pmatrix}
\tag{A.9}
$$

Appendix B

Implemented code

B.1 1D-hypoplasticity in FISH

```
set echo on ; sxx = zs11 ; syy = zs22 ; szz = zs33

; Date            Revision ; 10-June-02        First create

def m_hyperbo constitutive_model

f_prop ei sigmax f_prop c_zs22 float inc_zs22

case_of mode ; Initialization
    case 1

; Running section
    case 2
    zs11= 0.0
    zs22=zs22+ei*zde22-ei*abs(zde22)*zs22/sigmax
    zs33=zs33+0.0
    zs12=zs12+0.0

; Cummulative strain
    inc_zde22 = abs(zde22)
    if zsub > 0.0 then
    c_zde22   = c_zde22 + inc_zde22
;Do not divide by zsub because cum. strain will be reduced
    inc_zde22 = 0.0
    end_if

  ; Max modulus
    case 3
    cm_max=2e5
    sm_max=2e5
; cm_max=maximum confined modulus ; cm_max is used by FLAC to compute a
stable timestep ; it is essential that a value is returned for this variable
; for an Elastic model cm_max = K+(4G/3) ; sm_max=the shear modulus end_case
end
```

B.2 Hypoplasticity by Wu

```
// file: userhypo.cpp
// This file for Hypoplastic Constitutive Model
// Modified on: 09-Jan-2003
// 22-June-2004    : Add structure tensor
#include "userhypo.h" #include <math.h>
//variables used by all model objects. Hence only one copy is maintained for all objects
//
// Plasticity Indicators
//
// One static instance is neccessary as a part of internal registration
// process of the model with FLAC/FLAC3D static UserHypoModel
```

```
userhypomodel(true); UserHypoModel::UserHypoModel(bool bRegister)
           :ConstitutiveModel(mnUserHypoModel,bRegister),
           dC1(0.0), dC2(0.0), dC3(0.0), dC4(0.0) {
             }
const char **UserHypoModel::Properties(void) const {
  static const char *strKey[] =
  {
    "C1", "C2","C3","C4", 0
  };
  return(strKey);
}
 const char **UserHypoModel::States(void) const {
  static const char *strKey[] = {
    "shear-n","tension-n","shear-p","tension-p",0
  };
  return(strKey);
}
 /*  * Note: Maintain order of property input/output*/ double
UserHypoModel::GetProperty(unsigned ul) const {
  switch (ul)
  {
    case 1:  return(dC1);
    case 2:  return(dC2);
    case 3:  return(dC3);
    case 4:  return(dC4);
  }
  return(0.0);
}
void UserHypoModel::SetProperty(unsigned ul,const double &dVal) {
  switch (ul)
  {
    case 1: dC1 = dVal; break;
    case 2: dC2 = dVal; break;
    case 3: dC3 = dVal; break;
    case 4: dC4 = dVal; break;
  }
}
//Detects type mismatch error and returns error string. otherwise returns 0

const char *UserHypoModel::Copy(const ConstitutiveModel *cm) {
  const char *str = ConstitutiveModel::Copy(cm);
  if (str) return(str);
  UserHypoModel *mm = (UserHypoModel *)cm;
  dC1   = mm->dC1;
  dC2   = mm->dC2;
  dC3   = mm->dC3;
  dC4   = mm->dC4;
  return(0);
}
const char *UserHypoModel::Initialize(unsigned uDim,State *) {
  if ((uDim!=2)&&(uDim!=3)) return("Illegal dimension in UserHypo constitutive model");

//      ----- Deleted ------
  return(0);
}
const char *UserHypoModel::Run(unsigned uDim,State *ps) {
  if ((uDim!=3)&&(uDim!=2)) return("Illegal dimension in Hypo constitutive model");

// --- plasticity indicator: store 'now' info. as 'past' and turn 'now' info off --
  int iPlas = 0;

// --- trial elastic stresses ---  But I change to Hypoplastic equation!!


// get incremental stresses form stnS.   -- by using pointer *ps
// get incremental strian from stnE.     -- by using pointer *ps
  double dE11 = ps->stnE.d11;      // incremental strain de11
  double dE22 = ps->stnE.d22;
  double dE33 = ps->stnE.d33;
  double dE12 = ps->stnE.d12;

  double dS11 = ps->stnS.d11;      // incremntal stress dsig11
  double dS22 = ps->stnS.d22;
  double dS33 = ps->stnS.d33;
  double dS12 = ps->stnS.d12;

  double trT, trTD, normD, trTD_trT, normD_trT, trT_3;
  double dS11a, dS11b, dS11c, dS11d, incS11;
  double dS22a, dS22b, dS22c, dS22d, incS22;
  double dS33a, dS33b, dS33c, dS33d, incS33;
```

```
    double dS12a, dS12b, dS12c, dS12d, incS12;

// General varible for Hypoplastic con equa

    trT    = dS11 + dS22 + dS33;
    trTD   = dS11*dE11 + dS22*dE22 + dS33*dE33 + (2.0*dS12*dE12);
    normD  = sqrt((dE11*dE11) + (dE22*dE22) + (dE33*dE33) + (2.0*dE12*dE12));

    trTD_trT  = trTD/trT;
    normD_trT = normD/trT;
    trT_3     = trT/3.0;

// Get updated stresses from Hypoplastic

// ------ inc sig11
    dS11a = dC1 * trT * dE11;
    dS11b = dC2 * trTD_trT * dS11;
    dS11c = dC3 * normD_trT * ((dS11 * dS11) + (dS12 * dS12));
    dS11d = dC4 * normD_trT * ((dS11 - trT_3) * (dS11 - trT_3) + (dS12 * dS12));
    incS11= dS11a + dS11b + dS11c + dS11d;

// ------ inc sig22
    dS22a = dC1 * trT * dE22;
    dS22b = dC2 * trTD_trT * dS22;
    dS22c = dC3 * normD_trT * ((dS22 * dS22) + (dS12 * dS12));
    dS22d = dC4 * normD_trT * ((dS22 - trT_3) * (dS22 - trT_3) + (dS12 * dS12));
    incS22= dS22a + dS22b + dS22c + dS22d;

// ------ inc sig33
    dS33a = dC1 * trT * dE33;
    dS33b = dC2 * trTD_trT * dS33;
    dS33c = dC3 * normD_trT * (dS33 * dS33);
    dS33d = dC4 * normD_trT * (dS33 - trT_3)*(dS33 - trT_3);
    incS33= dS33a + dS33b + dS33c + dS33d;

// ------ inc sig12
    dS12a = dC1 * trT * dE12;
    dS12b = dC2 * trTD_trT * dS12;
    dS12c = dC3 * normD_trT * ((dS11 * dS12) + (dS22 * dS12));
    dS12d = dC4 * normD_trT * ((dS11 - trT_3)*dS12 + (dS22-trT_3)*dS12);
    incS12= dS12a + dS12b + dS12c + dS12d;


// **** Updated four stress (stnS.d__) by stress increment ****
    ps->stnS.d11 += incS11;       // += (dS11a + dS11b + dS11c + dS11d);
    ps->stnS.d22 += incS22;       // +=(dS22a + dS22b + dS22c + dS22d);
    ps->stnS.d33 += incS33;       // +=(dS33a + dS33b + dS33c + dS33d);
    ps->stnS.d12 += incS12;       // +=(dS12a + dS12b + dS12c + dS12d);
    ps->stnS.d13 += ps->stnE.d13;
    ps->stnS.d23 += ps->stnE.d23;
/* --- calculate and sort ps->incips->l stresses and ps->incips->l directions
--- */

    Axes aDir;
    double dPrinMin,dPrinMid,dPrinMax,sdif=0.0,psdif=0.0;
    int icase=0;

    bool bFast=ps->stnS.Resoltopris(&dPrinMin,&dPrinMid,&dPrinMax,
        &aDir,uDim, &icase, &sdif, &psdif);

    double dPrinMinCopy = dPrinMin;
    double dPrinMidCopy = dPrinMid;
    double dPrinMaxCopy = dPrinMax;

    if (iPlas)
    {
      ps->stnS.Resoltoglob(dPrinMin,dPrinMid, dPrinMax, aDir, dPrinMinCopy,
          dPrinMidCopy,dPrinMaxCopy, uDim, icase, sdif,  psdif, bFast);

      ps->bViscous = false; // Inhibit stiffness-damping terms
    }
    else
    {
      ps->bViscous = true;  // Allow stiffness-damping terms
    }
    return(0);
}


//  Save all properties for the model
```

```
//  Note: It is not necessary to save and restore member variables
//  that would be initialized. This reduces the size of save file.

// Checks for type mismatch and returns error string. Otherwise 0.

const char *UserHypoModel::SaveRestore(ModelSaveObject *mso) {
  const char *str = ConstitutiveModel::SaveRestore(mso);
  if (str) return(str);
  mso->Initialize(4,0);
  mso->Save(0,dC1);
  mso->Save(1,dC2);
  mso->Save(2,dC3);
  mso->Save(3,dC4);
  return(0);
}
// EOF
```

B.3 Hypoplasticity by von Wolffersdorff

```
// file: userhypovw.cpp
// Hypoplastic constitutive equation Version von Wolffersdorff
// ******************** Important Notes ************************
// 24 March 2003      : Revised the code
// 27 March 2003      : Add conmodulus to change max_confined modulus
// 06 Novem 2003      : Modify code to check correctness of implementation

#include <iostream.h> #include <math.h> #include <iomanip.h> #include
<fstream.h>

void main () { double Pi   = 3.14159265358979; double d1d3 = 1.0/3.0; double
d3d2 = 3.0/2.0;
// One static instance is neccessary as a part of internal registration
process of the model with FLAC/FLAC3D static UserHypovwModel
userhypovwmodel(true);

UserHypovwModel::UserHypovwModel(bool bRegister)
        :ConstitutiveModel(mnUserHypovwModel,bRegister),
         dhs(0.0), dn(0.0), dphic(0.0),
         dec0(0.0), ded0(0.0), dei0(0.0),
         dalpha(0.0), dbeta(0.0), dvoidz(0.0), conmodulus(0.0)
{ }

const char **UserHypovwModel::Properties(void) const {
  static const char *strKey[] =
  {
    "hs", "n","phic","ec0","ed0", "ei0", "alpha", "beta", "voidz", "conmodulus", 0
  };
  return(strKey);
}

const char **UserHypovwModel::States(void) const {
  static const char *strKey[] = {
    "shear-n","tension-n","shear-p","tension-p",0 // Not used in Hypoplastic law
  };
  return(strKey);
}

/*  * Note: Maintain order of property input/output*/ double
UserHypovwModel::GetProperty(unsigned ul) const {
  switch (ul)
  {
    case 1:  return(dhs);
    case 2:  return(dn);
    case 3:  return(dphic);
    case 4:  return(dec0);
    case 5:  return(ded0);
    case 6:  return(dei0);
    case 7:  return(dalpha);
    case 8:  return(dbeta);
    case 9:  return(dvoidz);
    case 10:  return(conmodulus);
  }
```

```
      return(0.0);
}

void UserHypovwModel::SetProperty(unsigned ul,const double &dVal) {
  switch (ul)
  {
    case 1: dhs       = dVal; break;
    case 2: dn        = dVal; break;
    case 3: dphic     = dVal; break;
    case 4: dec0      = dVal; break;
    case 5: ded0      = dVal; break;
    case 6: dei0      = dVal; break;
    case 7: dalpha    = dVal; break;
    case 8: dbeta     = dVal; break;
    case 9: dvoidz    = dVal; break;
    case 10:conmodulus = dVal; break;
  }
}


//Detects type mismatch error and returns error string. otherwise returns 0

const char *UserHypovwModel::Copy(const ConstitutiveModel *cm) {
  const char *str = ConstitutiveModel::Copy(cm);
  if (str) return(str);
  UserHypovwModel *mm = (UserHypovwModel *)cm;
  dhs       = mm->dhs;
  dn        = mm->dn;
  dphic     = mm->dphic;
  dec0      = mm->dec0;
  ded0      = mm->ded0;
  dei0      = mm->dei0;
  dalpha    = mm->dalpha;
  dbeta     = mm->dbeta;
  dvoidz    = mm->dvoidz;
  conmodulus= mm->conmodulus;
  return(0);
}

const char *UserHypovwModel::Initialize(unsigned uDim,State *) {
  if ((uDim!=2)&&(uDim!=3))
  return("Illegal dimension in UserHypo von Wolffersdorff constitutive model");

//      ----- Deleted ------

  return(0);
}

const char *UserHypovwModel::Run(unsigned uDim,State *ps) {
  if ((uDim!=3)&&(uDim!=2))
  return("Illegal dimension in Hypo von Wolffersdorff constitutive model");


  static double dSumepsvol;
  int iPlas = 0;
//*********************************************************************************
  if (ps ->bySubZone == 0)        // Initialize stacks to calculate hardening parameters for zone
  {
      dSumepsvol = 0.0;           // Accumulate void for the zone.
  }
//*********************************************************************************


// --- trial elastic stresses --- But I change to Hypoplastic equation!!

    double dE11 = ps->stnE.d11;      // let dEij store incremental strain tensor
    double dE22 = ps->stnE.d22;
    double dE33 = ps->stnE.d33;
    double dE12 = ps->stnE.d12;

    double dS11 = ps->stnS.d11;      // let dSij store stress tensor
    double dS22 = ps->stnS.d22;
    double dS33 = ps->stnS.d33;
    double dS12 = ps->stnS.d12;


// ---- General variables
    double trT, trTvTv, normTvs;
    double Tvs11, Tvs22, Tvs33, Tvs12, trTvsTvs_temp, trTvsTvs, trTvsTvsTvs;
```

```
    double exp_f;
    double ei, ec, ed;
    double phi_cr, a_f, fb_t, fb_b, fb, fe;
    double tan_psi, cos_3q;
    double X1, X2, X3, X4;

    trT = dS11 + dS22 + dS33;


    Tvs11 = dS11/trT - d1d3;
    Tvs22 = dS22/trT - d1d3;
    Tvs33 = dS33/trT - d1d3;
    Tvs12 = dS12/trT;

    exp_f = exp(-1*pow((-1*trT/dhs),dn));

    ei = dei0*exp_f;
    ec = dec0*exp_f;
    ed = ded0*exp_f;

    phi_cr = dphic*Pi/180.0;

    a_f = sqrt(3.0) * (3.0 - sin(phi_cr)) / (2.0*sqrt(2.0) * sin(phi_cr));

// ------- Start term 1 to get X1 ---------------
    fb_t = (dhs/dn)*((1+ei)/ei)*(pow((dei0/dec0),dbeta))*(pow((-1*trT/dhs),(1.0-dn)));
    fb_b = 3.0 + a_f*a_f - sqrt(3.0)*a_f*(pow(((dei0-ded0)/(dec0-ded0)),dalpha));
    fb   = fb_t/fb_b;

// using dvoidz here check it out!!!!!! voidz: current state void ratio
    fe    = pow(ec/dvoidz, dbeta);
    trTvTv = ((dS11*dS11) + (dS22*dS22) + (dS33*dS33) + (2.0*dS12*dS12))/(trT*trT);
    X1    = fb*fe/trTvTv;          // X1 : term 1

// ------- Start term 2 to get X2 ---------------
    double F_f1, F_f2, F_f;

    normTvs = sqrt(Tvs11*Tvs11 + Tvs22*Tvs22 + Tvs33*Tvs33 + 2.0*Tvs12*Tvs12);
    tan_psi = sqrt(3.0)*normTvs;
    trTvsTvs_temp  = Tvs11*Tvs11 + Tvs22*Tvs22 + Tvs33*Tvs33 + 2.0*Tvs12*Tvs12;

    if (trTvsTvs_temp == 0.0)
    {
        trTvsTvs = 1.0;
    }
    else
    {
        trTvsTvs = trTvsTvs_temp;
    }

    trTvsTvsTvs = pow(Tvs11,3) + pow(Tvs22,3) + pow(Tvs33,3) +(3.0*Tvs12*Tvs12*Tvs11)
    + (3.0*Tvs12*Tvs12*Tvs22);
    cos_3q = -1.0*sqrt(6.)*trTvsTvsTvs/pow(trTvsTvs,d3d2);
    F_f1 = sqrt((tan_psi*tan_psi/8.0) + ((2.0-tan_psi*tan_psi)/(2.0 + sqrt(2.0)*tan_psi*cos_3q)));
    F_f2 = tan_psi/(2.0*sqrt(2.0));
    F_f  = F_f1 - F_f2;
    X2   = F_f*F_f;                 // X2 : term 2

// ------- Start term 3 to get X3 -----------------
    double trTvD;

    trTvD = (dS11*dE11 + dS22*dE22 + dS33*dE33 + 2.0*dS12*dE12)/trT;
    X3    = (a_f*a_f)*trTvD;        // X3 : term 3

// ------- Start term 4 to get X4 ----------------
    double fd;
    double normD;

    fd    = pow(((dvoidz - ed)/(ec - ed)),dalpha); // change from dvoidini to dvoidz
    normD = sqrt(dE11*dE11 + dE22*dE22 + dE33*dE33 + 2.0*dE12*dE12);
    X4    = fd * a_f * F_f * normD;


//------- Start update zone stresses------

    double dS11i, dS22i, dS33i, dS12i;

    dS11i = X1*(X2*dE11 + X3*(dS11/trT) + X4*(2.0*dS11/trT - d1d3));
    dS22i = X1*(X2*dE22 + X3*(dS22/trT) + X4*(2.0*dS22/trT - d1d3));
    dS33i = X1*(X2*dE33 + X3*(dS33/trT) + X4*(2.0*dS33/trT - d1d3));
```

```
      dS12i = X1*(X2*dE12 + X3*(dS12/trT) + X4*(2.0*dS12/trT));

// **** Updated four stress (stnS.d__) by stress increment ****

      ps->stnS.d11 += dS11i;
      ps->stnS.d22 += dS22i;
      ps->stnS.d33 += dS33i;
      ps->stnS.d12 += dS12i;
      ps->stnS.d13 += ps->stnE.d13;
      ps->stnS.d23 += ps->stnE.d23;


// **** Accumulation and update volumetric strain ******
// remove accumulated volumetric strain

      dSumepsvol = dSumepsvol + (dE11+dE22+dE33)*(ps->dSubZoneVolume);

// if this is the final triangle -> averaging must be done.
      if (ps->bySubZone == ps->byTotSubZones-1)
      {
          double dAux = 1.0 / (ps->dZoneVolume);
          if (ps->byOverlay == 2) dAux *= 0.5;
          dvoidz = dvoidz + (1+dvoidz)*dSumepsvol*dAux;
      }

// ------------ End of Hypoplastic Law -------------------------------------

// --- calculate and sort ps->incips->l stresses and ps->incips->l directions ---

   Axes aDir;
   double dPrinMin,dPrinMid,dPrinMax,sdif=0.0,psdif=0.0;
   int icase=0;

   bool bFast=ps->stnS.Resoltopris(&dPrinMin,&dPrinMid,&dPrinMax,
       &aDir,uDim, &icase, &sdif, &psdif);

   double dPrinMinCopy = dPrinMin;
   double dPrinMidCopy = dPrinMid;
   double dPrinMaxCopy = dPrinMax;

   if (iPlas)
   {
     ps->stnS.Resoltoglob(dPrinMin,dPrinMid, dPrinMax, aDir, dPrinMinCopy,
         dPrinMidCopy,dPrinMaxCopy, uDim, icase, sdif,  psdif, bFast);

     ps->bViscous = false; // Inhibit stiffness-damping terms
   }
   else
   {
     ps->bViscous = true;  // Allow stiffness-damping terms
   }

   return(0);
}

//  Save all properties for the model
//  Note: It is not necessary to save and restore member variables
//  that would be initialized. This reduces the size of save file.
// Checks for type mismatch and returns error string. Otherwise 0.
const char *UserHypovwModel::SaveRestore(ModelSaveObject *mso) {
  const char *str = ConstitutiveModel::SaveRestore(mso);
  if (str) return(str);
  mso->Initialize(10,0);
  mso->Save(0,dhs);
  mso->Save(1,dn);
  mso->Save(2,dphic);
  mso->Save(3,dec0);
  mso->Save(4,ded0);
  mso->Save(5,dei0);
  mso->Save(6,dalpha);
  mso->Save(7,dbeta);
  mso->Save(8,dvoidz);
  mso->Save(9,conmodulus);
  return(0);
}
    // 9 represents 9 properties that are doubles
```

```
  // and 0 represents 0 properties that are integers
// EOF
```

B.4   Hypoplasticity with structure tensor

```
#include "userhypo.h" #include <math.h>

// One static instance is neccessary as a part of internal
//registration process of the model with FLAC/FLAC3D static
UserHypoModel userhypomodel(true);
UserHypoModel::UserHypoModel(bool bRegister)
         :ConstitutiveModel(mnUserHypoModel,bRegister),
          dC1(0.0), dC2(0.0), dC3(0.0), dC4(0.0) {
}
const char **UserHypoModel::Properties(void) const {
  static const char *strKey[] =
  {
    "C1", "C2","C3","C4", 0
  };
  return(strKey);
}
 const char **UserHypoModel::States(void) const {
  static const char *strKey[] = {
    "shear-n","tension-n","shear-p","tension-p",0
  };
  return(strKey);
}

/*  * Note: Maintain order of property input/output*/ double
UserHypoModel::GetProperty(unsigned ul) const {
  switch (ul)
  {
    case 1:  return(dC1);
    case 2:  return(dC2);
    case 3:  return(dC3);
    case 4:  return(dC4);
  }
  return(0.0);
}
 void UserHypoModel::SetProperty(unsigned ul,const double &dVal)
{
  switch (ul)
  {
    case 1: dC1 = dVal; break;
    case 2: dC2 = dVal; break;
    case 3: dC3 = dVal; break;
    case 4: dC4 = dVal; break;
  }
}
//Detects type mismatch error and returns error string. otherwise returns 0

const char *UserHypoModel::Copy(const ConstitutiveModel *cm) {
  const char *str = ConstitutiveModel::Copy(cm);
  if (str) return(str);
  UserHypoModel *mm = (UserHypoModel *)cm;
  dC1    = mm->dC1;
  dC2    = mm->dC2;
  dC3    = mm->dC3;
  dC4    = mm->dC4;
  return(0);
}
const char *UserHypoModel::Initialize(unsigned uDim,State *) {
  if ((uDim!=2)&&(uDim!=3)) return("Illegal dimension in UserHypo constitutive model");
  return(0);
}
const char *UserHypoModel::Run(unsigned uDim,State *ps) {
  if ((uDim!=3)&&(uDim!=2)) return("Illegal dimension in Hypo constitutive model");

// --- plasticity indicator: store 'now' info. as 'past' and turn 'now' info off --
  int iPlas = 0;
// --- trial elastic stresses ---  But I change to Hypoplastic equation!!
// get incremental stresses form stnS.   -- by using pointer *ps
// get incremental strian from stnE.     -- by using pointer *ps
  double dE11 = ps->stnE.d11;        // incremental strain de11
```

```
    double dE22 = ps->stnE.d22;
    double dE33 = ps->stnE.d33;
    double dE12 = ps->stnE.d12;

    double dS11 = ps->stnS.d11;        // incremntal stress dsig11
    double dS22 = ps->stnS.d22;
    double dS33 = ps->stnS.d33;
    double dS12 = ps->stnS.d12;

    double trT, trTD, normD, trTD_trT, normD_trT, trT_3;
    double dS11a, dS11b, dS11c, dS11d, incS11;
    double dS22a, dS22b, dS22c, dS22d, incS22;
    double dS33a, dS33b, dS33c, dS33d, incS33;
    double dS12a, dS12b, dS12c, dS12d, incS12;

// General varible for Hypoplastic con equa

    trT   = dS11 + dS22 + dS33;
    trTD  = dS11*dE11 + dS22*dE22 + dS33*dE33 + (2.0*dS12*dE12);
    normD = sqrt((dE11*dE11) + (dE22*dE22) + (dE33*dE33) + (2.0*dE12*dE12));

    trTD_trT  = trTD/trT;
    normD_trT = normD/trT;
    trT_3     = trT/3.0;

// Get updated stresses from Hypoplastic
// ------ inc sig11
    dS11a = dC1 * trT * dE11;
    dS11b = dC2 * trTD_trT * dS11;
    dS11c = dC3 * normD_trT * ((dS11 * dS11) + (dS12 * dS12));
    dS11d = dC4 * normD_trT * ((dS11 - trT_3) * (dS11 - trT_3) + (dS12 * dS12));
    incS11= dS11a + dS11b + dS11c + dS11d;
// ------ inc sig22
    dS22a = dC1 * trT * dE22;
    dS22b = dC2 * trTD_trT * dS22;
    dS22c = dC3 * normD_trT * ((dS22 * dS22) + (dS12 * dS12));
    dS22d = dC4 * normD_trT * ((dS22 - trT_3) * (dS22 - trT_3) + (dS12 * dS12));
    incS22= dS22a + dS22b + dS22c + dS22d;
// ------ inc sig33
    dS33a = dC1 * trT * dE33;
    dS33b = dC2 * trTD_trT * dS33;
    dS33c = dC3 * normD_trT * (dS33 * dS33);
    dS33d = dC4 * normD_trT * (dS33 - trT_3)*(dS33 - trT_3);
    incS33= dS33a + dS33b + dS33c + dS33d;
// ------ inc sig12
    dS12a = dC1 * trT * dE12;
    dS12b = dC2 * trTD_trT * dS12;
    dS12c = dC3 * normD_trT * ((dS11 * dS12) + (dS22 * dS12));
    dS12d = dC4 * normD_trT * ((dS11 - trT_3)*dS12 + (dS22-trT_3)*dS12);
    incS12= dS12a + dS12b + dS12c + dS12d;
// **** Updated four stress (stnS.d__) by stress increment ****
    ps->stnS.d11 += incS11;
    ps->stnS.d22 += incS22;
    ps->stnS.d33 += incS33;
    ps->stnS.d12 += incS12;
    ps->stnS.d13 += ps->stnE.d13;
    ps->stnS.d23 += ps->stnE.d23;
    /* --- calculate and sort ps->incips->l stresses and ps->incips->l directions --- */
    Axes aDir;
    double dPrinMin,dPrinMid,dPrinMax,sdif=0.0,psdif=0.0;
    int icase=0;
    bool bFast=ps->stnS.Resoltopris(&dPrinMin,&dPrinMid,&dPrinMax,
        &aDir,uDim, &icase, &sdif, &psdif);

    double dPrinMinCopy = dPrinMin;
    double dPrinMidCopy = dPrinMid;
    double dPrinMaxCopy = dPrinMax;
    if (iPlas)
    {
      ps->stnS.Resoltoglob(dPrinMin,dPrinMid, dPrinMax, aDir, dPrinMinCopy,
          dPrinMidCopy,dPrinMaxCopy, uDim, icase, sdif,  psdif, bFast);

      ps->bViscous = false; // Inhibit stiffness-damping terms
    }
    else
    {
      ps->bViscous = true;  // Allow stiffness-damping terms
    }
    return(0);
}
```

```
//  Save all properties for the model
//  Note: It is not necessary to save and restore member variables
//  that would be initialized. This reduces the size of save file.
// Checks for type mismatch and returns error string. Otherwise 0.

const char *UserHypoModel::SaveRestore(ModelSaveObject *mso) {
  const char *str = ConstitutiveModel::SaveRestore(mso);
  if (str) return(str);
  mso->Initialize(4,0);
  mso->Save(0,dC1);
  mso->Save(1,dC2);
  mso->Save(2,dC3);
  mso->Save(3,dC4);
  return(0);
}
// EOF
```